THE GREAT DEEP

JAMES HAMILTON-PATERSON

THE GREAT DEEP

THE SEA AND ITS THRESHOLDS

AN OWL BOOK

HENRY HOLT AND COMPANY
NEW YORK

Henry Holt and Company, Inc.
Publishers since 1866
115 West 18th Street
New York, New York 10011

Henry Holt® is a registered trademark
of Henry Holt and Company, Inc.

Library of Congress Cataloging-in-Publication Data
Hamilton-Paterson, James.
The great deep : the sea and its thresholds / James Hamilton-Paterson.
p.   cm.
Includes index.
1. Oceanography.   I. Title.
GC26.H36   1994   93–9035
551.46–dc20   CIP

ISBN 0-8050-2776-9

Henry Holt books are available for special promotions and
premiums. For details contact: Director, Special Markets.

First published in hardcover by
Random Century, London, in 1992.

First American edition published by Random House, Inc.

First Owl Book Edition—1993

Designed by J. K. Lambert

Printed in the United States of America
All first editions are printed on acid-free paper.∞
10 9 8 7 6 5 4 3 2 1

For my mother

Also

to the memory of

Ben Chong and Arnel Julao,

last seen on 20 December 1987

hoping to stow away aboard M/V *Doña Paz*

in order to join their families

for Christmas

ACKNOWLEDGMENTS

I gratefully acknowledge the help given me in the writing of this book by the following:

Anna Badini, Anabel Briggs, Paul Brown, Peter Chaplin, Mario and Roberta D'Itria, Frank Donn, Earthtrust (Hawaii), the scientists and crew of the R/V *Farnella*, Stephen Feuchtwang, Bill Foster, Luis Go, Abdurrahim Hasim, Ellis Hillman, Frank Sionil José, Abdurahim Kenoh, Cdr. Reuben S. Lista (Philippine Navy), George May, James May, John May, Anita McConnell, Gilbert Tait, Saladin S. Teo, Marcello Vanni, Ian West.

I would like to thank the Institute of Oceanographic Sciences, Deacon Laboratory, and especially Mike Somers and Tony Rice for their courtesy and help. Above all, Quentin Huggett has been unfailingly patient and generous with his time, a prodigality which it is a pleasure to acknowledge.

My agents, Margaret and Andrew Hewson, deserve—and get—my very warmest thanks for their tireless support and help, as does Patricia Reynolds.

This book's commissioning editor was Richard Cohen, and I am conscious that it owes everything to his original encouragement. To that extent it remains his. The task of taking it on fell to Neil Belton, my present editor. I could not have wished for a more sympathetic and astute interpreter of a project whose intentions must initially have appeared idiosyncratic and opaque.

Finally, my gratitude to Mark Cousins is not easily expressed. One hardly thanks a friend for friendship, but one might well wish to put on record the affectionate recognition of a great debt. The entire book was written very far from his scrutiny, but scarcely a line without awareness of his presence. Private congruences apart, there was for me an especial punctuality about his public lectures at the Architectural Association in London, 1990–91, which had a profound effect on this book, just as the lectures had on all who heard them.

The above persons are exonerated from all responsibility for any errors of fact, judgment and taste, which are entirely my own.

ILLUSTRATION ACKNOWLEDGMENTS

I   from: *Zetetic Astronomy* (1873) 2nd edition © British Library, London

II   © British Museum, London

III   (i) & (ii) from: *The Structure and Distribution of Coral Reefs* by Charles Darwin (1889)
      (iii) & (iv) from: *Corals and Coral Islands* by James Dana (1872)

IV   from: *The Christian Miscellany* (1860)
      © Mary Evans Picture Library

V   from: *Twenty Thousand Leagues under the Sea* by Jules Verne
      © Mary Evans Picture Library

VI   *Miraculous Draught of Fishes* by Gustav Doré
      © Mary Evans Picture Library

VII   from: *Concise Oxford Atlas* (1952)
      © Oxford University Press

C O N T E N T S

THE GREAT DEEP

The world according to zetetics: not a globe but a disk of
geographical features bounded by a plane of infinite—or at
least unknowable—extension and thickness. The hatched
outer ring, all of which represents South, marks the insur-
mountable Great Ice Barrier.

I am lost. . . . These are the words the swimmer addresses in panic to the sunny universe into which he emerges, blowing water, disoriented. Ten minutes ago, perhaps twenty, he had set some fishing lines and slipped over the side of his tiny craft—a wooden insect with two bamboo outriggers— with a cord tying its prow to his ankle. He had been lying face down in the ocean, sun on his back, staring through the first thirty of a thousand meters of water. In the tropics these upper waters are flooded with light. Bright spicules drift past his eyes, crimson and electric blue, the jeweled phytoplankton streaming about the globe performing infinitesimal acts of chemistry which, much multiplied, succor all earthly life. By swimming down a couple of dozen feet he can look up and view other creatures from below: a shoal of garfish (whose bones are bright green) so high up their backs graze the rumpled mirror of air, the occasional flying fish breaking out and vanishing. The swimmer reflects on this mirror, imagining the sky weighing down on the sea and the sea holding up the atmosphere, curious about what exactly can be happening at the interface. If it were possible to magnify the activity, surely a buzzing skin of molecules would be revealed? Water molecules and air molecules so intermixed and satu-rated with atoms in common it would be undecidable which medium they constituted. At what point did these milling particles become waves? The swimmer loses himself in this quantum pun, in his speculations about boundaries, then suddenly an awareness breaks in that something is missing. There is a steady tug at his ankle, but too light. The long cord trails downward, still firmly knotted to one foot. It is the boat which has gone.

His first act of panic is to spin in the water while trying to stand up in it: once, twice, three times, quartering a featureless horizon. Nothing. He is anchored by twenty feet of thin abaca hemp to the ocean. His masked face rams back through the surface as if by a miracle of misplacement he might discover the boat floating at ease in a fourth dimension some fathoms below. Nothing. The cord hangs down like the corals called sea whips, slightly kinked, whiskers of fiber standing out clear in this awesome lens right to the bobble of the knot in its end. The word this knot transmits through the water is "adrift."

The swimmer jerks his head up into the air again. Everything is plain. It is not possible, yet the boat has gone. I am lost. Panicked, he pants and spins, boatless, landless, and with the visceral ache of pure fear at what he has abruptly become: all alone and floating in the Pacific Ocean. Reason attempts to be reasonable. How far away could a boat possibly drift on a windless day? Also, eye level is barely six inches above sea level; a boat whose freeboard is little more than three times that could easily be hidden by the least swell. It is no doubt bobbing in and out of visibility even as he happens to scan the wrong horizon. . . . In any case, something altogether calmer is taking over: a lassitude, a fatalism whose roots reach back not to the beginning of his own life but, like the rope on his ankle, down into the sea itself. The twisted fibers, like ancient strands of DNA, connect him with vanished deeps, to primordial oceans lying in different beds. If he is lost now it is because he was already lost before ever setting foot in a boat, before infancy itself. He has no proper existence at all, being only a tiny hole in the water shaped like the lower two thirds of a man. There is no way the tons of ocean can be held apart and prevented from filling the mold.

Yet it is not possible to give up, to go within minutes from being fit and happily occupied to renouncing life as if fatally injured. Fear returns in cycles. Looking around the liquid wastes beneath a brilliant sky he is set upon at intervals by the adrenal thought: This cannot be happening to me. . . . But it is. Then for a little while it is not; and in between assessing his chances of death by drowning, shark attack or exposure, into the swimmer's mind comes a sharp, vainglorious image of his predicament. Lacking all coordinates, he sees his own head occupying a fixed place. He pictures it sticking out of that expanse of curved blue ocean, a little round ball burnished with sun like the brass knob on top of a school globe. In his moment of loss he becomes the pivotal point about which the entire Earth turns.

I

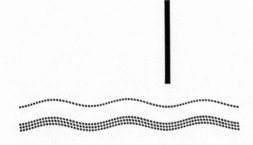

CHARTS
AND NAMING

1. CHARTS AND NAMING

They are a blank to most people, these bald oceans, much as they appear on a globe. As such they have their own coherence as two-dimensional representations of *not-land*. "The sea" survives elsewhere in piecemeal images, scattered pictures which never link up. They include beach scenes from summer holidays, an aunt being seasick, storms, sunken treasure, pirates, monsters of the deep. . . . "The sea" as a reservoir of private imagery and public myth remains on the one hand without limits, on the other banally circumscribed. Yet it haunts us. We are full of its beauty, of that strange power it gives off which echoes through our racial history and fills our language with its metaphors. The salt which is in seawater is in our blood and tears and sweat. The lungs of an infant *in utero* can be seen rhythmically breathing as it inhales and expels amniotic fluid, even as its oxygen supply comes from the mother's bloodstream via the umbilicus. Each of us has breathed warm saline for days on end and survived. The lungs them-selves derive from fused pharyngeal pouches, and branchial clefts ("gill slits") still form temporarily in all chordate embryos, including humans, reminding us that something which became *Homo* did crawl up a beach many millions of years ago. The satisfaction for certain people of walking back down a beach and into the sea is akin to that of a long-postponed homecoming.

Too late, though. We have lost our place and no longer know how to return. It is never quite enough, the ecstasy of splashing or the torpid floating while a fine slick of tanning oil spreads its iridescent pollution around us. Too much knowledge, too much strangeness slips by. Maybe we should approach the whole thing through science. Marine biology will tell us about the phytoplankton in their gem-like flakes, trigonome-try will reveal where we are. The business of orientation on the blank and shifting waters of the open sea, of establishing a fixed *point de repère* which is not a landmark, is central not merely to navigation but to various sciences coming under the general heading of oceanography. (The very word suggests a difficulty, the writing down of an ocean.) Until a certain moment in history there must have been a conceptual impossibility in the idea of a sea chart without a coastline on it, implying as it logically would the drawing of lines and boundaries—albeit notional ones—on a fluid surface. Being boatless and lost in mid-ocean at noon in the tropics at least makes vivid certain problems which have faced all navigators and cartographers, with the sun directly overhead and the seabed far out of sight a mile below. The panic of a careless swimmer keen to avoid joining his ancestors thus makes a good starting point as he twirls despairingly round and round like a demented compass needle in search of any bearing, any point of reference, any direction other than down.

The swimmer wanted to know many things. Gazing straight down beyond his parsnip-white feet into the purple depths he wanted to know not just what was down there but how it might be mapped. How did one make an accurate chart of the seven tenths of the Earth's crust hidden from sight? Since he had always wanted to know such things he wondered why he had never become a marine scientist. This was impon-derable, so he inverted the question and speculated about what it is that makes a scientist choose his particular field of study. This swimmer, who is almost as obsessed with the idea of the sea as with its actuality, wondered if oceanographers necessarily shared his own obsession, and if so, to what effect. Did they love to go delving into the sea? Was it scientifically useful to be imaginatively caught up in the deeps?

~

Such speculations bring me in due time (early December 1990) to Pier 40 of Honolulu Port and aboard the R/V *Farnella*, a British-registered research vessel. The *Farnella* is under charter to the United States Geological Survey, which is using her to map the seabed contained by the US 200-mile Exclusive Economic Zone. It seems curious that this grand national task should be carried out using a system developed in Surrey and mounted aboard a converted Hull trawler. Some years ago the ship's refrigerators were taken out and her forward hold filled with huge spools of cable for towing instruments. The stern hold is now largely given over to a spacious air-conditioned laboratory humming with bays of computers, sonographs, plotters and assorted electronic gear. When we leave Honolulu we have aboard fourteen crew, thirteen scientists (American and British) and myself, my capacity officially listed as "Fly on wall."

We spend the first two days at sea simply reaching the area of the survey, which is that surrounding Johnston Island. At this moment Johnston Island is a high-risk, high-security area, since this is the atoll where the US is destroying its stocks of nerve-gas shells and other chemical weapons from bases in Germany. Several of the scientists had put ashore there briefly on a previous trip. "Can't land if you've got a beard," said one, "or even a couple of days' growth. Your gas mask mightn't be airtight." "Don't worry," said another. "We'll be over a hundred miles away most of the time and with any luck upwind."

I wander the ship, stare at the ocean rolling past, meet the captain. "He's temporary," a crew member tells me. "Wicked sense of humor." A dapper man, he had distinguished himself during the last Cod War of 1975 by deliberately ramming an Icelandic trawler. Otherwise, his sense of humor seems not greatly in evidence. Ah, but a previous captain, now, he'd been gross in all sorts of ways. "Pig fat, he was. Also, I remember up on the bridge, you'd never know when, sort of absentminded-like he'd take out his teeth to scratch the eczema on his legs, the top plate usually, and you'd see this dust like cornflour come puffing out of his trousers. Then he'd slippem back innagain." The sea has always been kind to eccentrics, and indeed the *Farnella*'s crew give off a feeling of dourly amiable tolerance, even toward scientists and flies on walls.

The main piece of equipment for bathymetry aboard, and the chief tool of the entire USGS program, is GLORIA: Geological Long Range Inclined Asdic ("asdic" being the British equivalent of the American "sonar"). Since seawater is four-and-a-half times better than air at conducting sound, some kind of sonar system remains the most effective way of mapping a seabed which may be anything up to eleven kilometers below the surface. The very earliest sonars were simple echo rangers. The sinking of the *Titanic* in 1912 had made finding some sort of defense against icebergs an urgent matter, but it was actually World War I—and specifically the development of submarine warfare—which encouraged sonar technology and in so doing gave the biggest boost to modern oceanography. One offshoot of military research was an efficient echo sounder for cable laying.

In 1854 Lieutenant Matthew Fontaine Maury of the US Depot of Charts and Instruments had published a profile of the Atlantic seabed, his *Bathymetrical Map of the North Atlantic Basin with Contour Lines Drawn in at 1,000, 2,000, 3,000 and 4,000 Fathoms*. This was based on only a couple of hundred deep soundings achieved with a weighted line, and its imposing title suggested a survey rather more thorough than it actually was. Nevertheless, it interested a good many people, among them a rich industrialist, Cyrus W. Field. Field had long dreamed of linking America with Europe by telegraphy. The seabed between New-foundland and Ireland was, according to Maury's survey, mostly plateau. Suddenly, the romantic notion of voices emerging from the deep took on real commercial possibility. In 1856 the Atlantic Telegraph Company was founded and two years later it laid the first transatlantic cable. Unfortunately, Maury's "plateau" turned out to include over 1,000 miles of fracture zone: the spine, in fact, of the Earth's largest mountain range, the Mid-Atlantic Ridge. This runs for 7,000 miles with peaks rising two-and-a-half miles above the seabed. Within three months the cable had broken. Under the continued pressure of the Company's commercial determination, though, Maury's team were inspired to de-velop better sounding devices and bathygraphy. The result was a perma-nent telegraph link, finally laid in 1866. Yet even the best sounding techniques using plummets and lines had great limitations (see Chapter

VI). At least another half-century was to elapse before the novel alternative of sonar was devised.

In 1922 a historic moment in cartography and oceanography occurred when the USS *Stewart* used a Navy Sonic Depth Finder to draw a continuous profile across the bed of the Atlantic, sixty-eight years after Maury had published his map. But whereas Maury's had involved large amounts of guesswork, the *Stewart*'s involved none, aside from certain problems of interpretation. Otherwise the soundings were fast, accurate and simple to take. The principle of sonar is straightforward. A pulse of sound is bounced off the sea floor and received by the vessel which sent it. Having made small allowances for such things as the speed of the vessel, it is a matter of the simplest arithmetic: Divide the time elapsed by two and multiply the result by 4,800, which gives the depth in feet, 4,800 feet per second being the mean speed of sound in water. (The layman is astonished only that, with a principle as thoroughly understood and long turned to practical use, it should have taken so long to develop radar, that apparently magical World War II conceptual breakthrough. Since then, with a good deal of technical wizardry but zero intellectual daring, radio pulses and even laser beams have been bounced off the Moon, ranging it to a matter of a few centimeters.)

There are, however, certain difficulties to this kind of bathymetric depth sounding. One can be intuited from the phrase "mean speed of sound in water," since in practice the speed of sound in seawater varies. It travels faster, for instance, if the temperature, pressure or salinity increases. This is to leave aside for the moment all the difficulties in interpreting the returning signals, which may be severely affected by shoals of fish and other powerful sources of scattering. A further drawback is that a series of pulses focused vertically downward beneath a ship's keel will give only a series of depths. In order to build up a contour map of how the sea floor actually looks, a ship using this method would have to make a laborious succession of very closely spaced passes. Otherwise, only a narrow cross-section of the seabed being traversed is possible.

A way around this difficulty was pioneered at the Institute of Oceanographic Sciences in Surrey. Instead of using vertical echo sounding, they developed a sidescan sonar in which the sound pulses are emitted sideways in two slanting fans, one on either side of the ship and at right angles to its course. The outer edges of these fans brush irregularities on the seabed as far away as thirty kilometers. The returning echoes, since they do not come merely from a single point directly beneath the vessel's keel, are complex and take skill and experience to interpret. Yet the profile which emerges from the plotters provides by far the clearest imagery yet possible of a continuous swath of seabed, certainly on a cost-effective scale. GLORIA's efficiency is striking. In good conditions and with the ship traveling at eight or more knots it can reveal a strip of seabed sixty kilometers wide. In twenty-four hours this amounts to mapping more than 20,000 km², or an area the size of Wales, in a day.

This cruise in the *Farnella* is one of a series at the end of a great project on the part of the US to map the 200-mile EEZ around all 19,924 km of its coastline. The EEZ of the continental US has already been mapped, including Alaska and the Aleutian Islands. Now *Farnella* is working away at the Hawaiian chain. As the scientists aboard keep telling me, only the GLORIA system could have covered such an area, and even that has taken ten years.

Such things are explained as the ship bounces and judders through seasonal Pacific rollers on her way to the survey area. On and off a tropic sun blazes the sea to a luminous indigo across which flying fish skip and glide. In the intermittent bursts of sunlight there is a stampede of scientists up onto the deck where they grill themselves on towels, redolent of coconut oil. December! Hawaii! Field trips sometimes throw in for free the costly ingredients of other people's holidays. They are watched by two boobies clumsily slithering on the button atop the radar mast while a small albatross flies elegant, mournful rings around *Farnella*. On successive days of watching I never once saw it move its wings, only soar and tilt in any direction and at any speed into the headlong breeze while keeping pace with the ship. It looked as though it knew by heart a map of the ocean's surface, a map no man will ever make.

Time was spent checking the various instruments which were soon to be lowered into the sea and towed behind us. First of all GLORIA itself: a large yellow torpedo lined inside with banks of transponders precisely angled to give the correct fan-shaped pulse. It sits in a hydraulic cradle directly over the ship's stern. From time to time technicians climb up to tighten a nut and pat it protectively. Even without its cable it is worth nearly half a million pounds. There is spare cable but only one GLORIA aboard. "We don't even like to think about that," is the response to the obvious "What if?"

Amidships, two little fish-like fat yellow bombs wait on wooden trestles. These are the 10 kilohertz and 3.5 kilohertz sonars. The first will act more or less like an old-fashioned depth sounder to give the distance to the seabed immediately beneath us. The second will send pulses up to 400 meters into the ocean floor to provide some idea of the underlying geology. It is explained that its readout, taken by itself, would be fairly meaningless, since in physics, as elsewhere, one cannot get something for nothing and the deeper penetration gained by using a higher frequency is at the expense of detail and spread. But taken in conjunction with information coming from other instruments these data will be usefully corroborative.

Over the starboard stern will go the air gun. This is a heavy pot of machined steel, something like the body of a pneumatic drill and similar in principle. Connected to a high-pressure air hose it will "fire" itself every ten seconds with a loud detonation and send soundwaves capable of penetrating up to one kilometer into the ocean floor. This is the oceanographer's equivalent of seismologists' "thumper trucks" which used to drive around the deserts of the Middle East, dropping huge blocks of concrete and listening for signs of oil-bearing strata. The air gun is a safer and simpler alternative to throwing overboard dynamite charges at timed intervals. Its echoes are received on a hydrophone "streamer," hundreds of meters of transparent plastic tube filled with bundles of multicolored sensor wires and light oil. The streamer may be towed well over a kilometer behind the ship and is of all the equipment the most susceptible to damage. (At the end of the cruise when the streamer is wound aboard there is a section leaking oil with a broken shark's tooth embedded in the gash.)

Over the port stern will go a magnetometer to measure magnetic variability in the Earth's crust, and down in the lab is a gravimeter to record differences in its gravitational field. This machine looks, and is, expensive. It is suspended in a cradle mounted in computer-controlled gimbals, dipping and tilting so it appears to be the one thing in the lab which is constantly in motion, whereas it is really the only thing aboard remaining utterly still while the ship gyrates about it. At supper the conversation turns to where might be the best place on Earth for setting high-jump records, a particular spot with significantly weaker gravity. All the best ones seem to be covered by a couple of miles of water. In response to a remark of mine which betrays real ignorance about gravity, Roger says kindly:

"I suppose one always imagines the surface of the oceans as basically flat. Ignoring waves and local storms, of course—they're just "noise." But apart from its being curved to fit the surface of the globe, one thinks of the sea as having to be flat because at school we're told water always finds its own level so as to be perfectly horizontal. On a small scale that's pretty much true, though when I was about ten I remember being surprised when someone pointed out that all rivers are tilted, and if you row upstream you're also rowing uphill as well as against the current. Anyway, since gravity varies from place to place it acts variably on the sea, too. When you start using instruments like the ones aboard this ship you really appreciate how the ocean surface actually dips and bulges all over the place. It shows up best from space."

He explains that by having enough satellites in orbit making passes over the same area, day after day for months on end, it was possible to build up a mean reading for the height of the sea's surface at that spot. It took a long time because there was a good deal of "noise" to be discounted: wind heaping, sudden areas of low atmospheric pressure which could suck the sea upward as if beneath a diaphragm, even very low-frequency waves with swells so long they might take half a day to pass. But if the satellites went on measuring the same spot for long enough, such fluctuations would even out and a geodetic point be established: a mean distance to the sea's surface as measured from the center of the Earth. By building up enough geodetic points it soon became clear that the oceans were anything but flat.

"What's more, if you match this up with the underlying features on the seabed, you'll find that the surface of the sea broadly mimics the topography underneath. And the reason for *that* is fluctuations in gravity, which depends on the density of the crustal material."

It is a pretty notion, that the sea follows the Earth's crust like a quilt laid over a lumpy mattress. It is also odd to think that to some extent the depths of the oceans can be read from space. Over a plate of steak-and-kidney pie (the galley makes no concessions to the tropics) I presume this means that sometimes a ship has to go uphill and downhill.

"Certainly. But the 'hills' are so slight you'd never know. We're talking a few tens of centimeters here, spread out over kilometers. Sure, a ship often has to go uphill, though it won't be using any more energy. All points on the hillock's surface have the same gravitational potential, obviously."

This is not obvious to me, and nor is it any more so after Roger has explained it several times in different ways. I tell myself that physics is humiliating not when it defeats the intellect but when it confounds the imagination. This makes me feel better. Giving up on me, he reverts to a sort of "Ripley's Believe-It-or-Not" mode suitable for lay company. Roger is himself a geologist and in describing the Planet gives the impression of talking about a beach ball underinflated with water: labile, plastic, sagging and crinkling and bulging. It is not only the oceans which respond tidally to the Sun and the Moon; the Earth's surface does as well, rising and falling twice a day. When the Moon is directly overhead it is pulled up by half a meter.[1] What is more, this elastic crust seems to have a frequency of its own at which it resonates. A Russian geologist, S. L. Soloviev of the Moscow Institute of Oceanology, recently made seismograms of micro-earthquakes under the Tyrrhenian Sea. Using bottom seismographs (developed from nuclear-explosion detectors originally designed to enforce the Test Ban Treaty), Soloviev began picking up a distinct, ultra-low-frequency oscillation which he thought was most likely the fundamental frequency of the Earth's crust itself.

That night I go to bed with my head full of marvels. In the course of the evening I had also learned that the sea levels at either end of the Panama Canal were different by nearly half a meter, and the same went for the sea on either side of the Florida peninsula. This was caused by things such as the heaping effect of wind and the Coriolis force. But I am most captivated by the idea of the Earth's crust vibrating at an ascertainable frequency, since it would theoretically be possible to calculate the precise note. True, it would probably not be a pure tone because there would be all sorts of harmonic interference from irregularities such

[1]The extremest pull is, of course, when the Sun and Moon are perfectly aligned, as in the total solar eclipse of July 1991. Astronomers observing this phenomenon from the top of Mauna Kea, an extinct volcano in Hawaii, noticed a minor eruption in nearby Mauna Loa, believed until then to be equally extinct. This is considered highly suggestive of there being "tides" in the Earth's crust as well as in its seas.

as mountain ranges. Yet it ought to be possible to determine the fundamental note of the Planet, the music of our spheroid.

I also wonder at the notion of the sea's surface modeling the plains
and mountains, chasms and basins beneath the keel. It is not hard to
believe this at the moment since we are all being thrown about our
bunks by the *Farnella*'s plungings as if she were plowing across rough
country. We have reached the foothills of the Necker Ridge. More than
two miles below us a mountain chain thrusts steeply upward. I bang
about in my wooden trough and tell myself this is just "noise."

Next morning the sea is quieter. The ship heaves to and a certain
tension comes over the scientists as one by one the precious instruments
are carefully deployed. Launchings and retrievals are the moments when
damage is most likely and, although there are several workshops aboard
for mechanical and electronic repairs, our sailing patterns are planned to
the nearest nautical mile for the next several days. Worries revolve
around personal responsibility for the correct functioning of machines.
The specter of disgrace and delay flits about the ship until everything is
safely in the water and the test readings are monitored. In the end these
anxieties are probably rooted less in codes of professionalism than in the
huge expense of modern oceanography. Great sums of money are being
lowered delicately into the ocean. Any delays would be tantamount to
damage, chunks of money becoming dislodged and drifting down to the
seabed, where they would dissolve and be lost forever. Even the crew
seems less laconic while all this is going on. When we are under way
again there is a feeling that the *Farnella* is more in the hands of scientists
than seamen and the crew can now be found in odd corners reading
copies of the Hull *Daily Mail* which were flown out in bundles to the
shipping agent in Honolulu. Shipboard life settles into routine. It is still
curious to be in the Pacific in a British ex-trawler with a television in the
lounge showing videocassettes of highlights from last season's Hull
Kingston Rovers matches. Not to mention the cuisine. At the same time
we are, as an IOS zealot proudly says, "at the leading edge of geophysical seabed surveying."

~

The issue of who is really in command of the ship is interesting, as is the
whole idea of a joint survey paid for by the US government using a
significant proportion of British equipment and scientists. In a legal

sense the captain has full and final responsibility for the ship. Yet it soon becomes plain that his actions are largely determined by the exigencies of the survey, which is costing the American taxpayer such a pretty penny. It is the USGS which has chartered the vessel and hired GLORIA and so calls the tune. On the other hand, the scientist formally in charge of this particular cruise is a Briton, one of GLORIA's original developers at IOS. At the same time, one of the young American women aboard is responsible to her government for completing this leg of the survey. . . . All this interweaving of authority is glossed for me as "A joint effort. Absolute cooperation and consultation. Democracy in action like you wouldn't believe." This is emphatically not science for the sake of science, a matter of drifting about the Pacific like the old *Challenger* in the 1870s, sounding here and dredging there at whim. This is time-and-motion science, with a given area of blank map to be filled in a given time. And the whole issue, for very cogent reasons of physics, hinges around the matter of navigation.

As is all too clear to anybody swimming in circles looking for a lost boat in the middle of the ocean, one has no position in water. When mapping the seabed from a moving ship, therefore, accurate navigation is of crucial importance. Without the ship's position being known from one second to the next the most beautiful chart of peaks, ravines and plateaus would be useless. The only thing known would be that they were down there somewhere. Establishing the ship's course along lines as straight as possible (always allowing for the Earth's curvature) requires much work, not least because the swaths GLORIA maps must lie next to each other without gaps or wasteful overlapping. On the chart table in the lab is the dot of Johnston Island, a pencil circle whose diameter represents 400 nautical miles inscribed about it. High up in its top left-hand quarter a chord shows the first leg we have just started. Next to it is written the estimated time at which we should come about for the return pass, each leg getting longer as we eat downward into the circle. If all goes well, by the end of a fortnight we should have hatched off about a quarter of the total area.

While the lab computers flicker with the instruments' returning signals, various repeater gauges give the ship's speed through the water, its speed over the ground, the wind speed and any consequent degree of yaw. If to remain on a straight course against a quartering wind and current the *Farnella* needs to sail crabwise, GLORIA's angle will also be fractionally oblique to its correct path. The result is that its signals will no longer be exactly at right angles to this course and the map will be distorted. Information on all these factors is fed into the computers, which correct for them. In order to determine the ship's position at any moment the *Farnella* uses GPS, or Global Positioning System. This depends on satellites and eventually, provided there are still spare slots in an already overcrowded geostationary orbit, the system will cover the Earth and in theory allow a person anywhere on the Planet's surface to determine his position to within a few meters. This would not be of the slightest use to a lost swimmer looking for his boat.

Bored with the sight of bright red digital figures flickering their decimal points on display panels I wander off in search of sound. Down in the forward hold, above the banging of the ship's forefoot into wave troughs, the chaffinch-like chinking of the ten kilohertz "fish" can be heard through the steel hull. Up on the stern deck there is a sharp cracking sound every ten seconds, the higher frequencies of the air gun's detonations being transmitted back up the compressed air hose. In the water astern white puddles dimple and churn to mark the boilings of released air. They follow the ship with the measured pace of footsteps.

Very occasionally from a chance position down in the hull, at some freak acoustical window, it is possible to hear GLORIA's peculiar yodel. The instrument emits a correlation signal; instead of a single bleep its pulses take the shape of a whistle which swoops up and down. This is so the echo will be unmistakable, the electronic ears listening for its return being tuned to exclude all other signals. Even so, knowing how to read the GLORIA trace as it emerges from the plotter is a matter of much experience. Since parts of the signal are making a round journey of sixty kilometers or so, while others may travel only five (that edge of the fan nearest the ship), the returning echoes become mixed up with the fresh outgoing pulses, even with still fainter returns from previous signals. There may also be leakage and scattering, with stray echoes reflected back down from the water's surface.

How different the *Farnella* is from the old *Challenger*! The real distinction between this kind of oceanography and all that went before is not merely that the technology has changed, and with it the techniques for analyzing data. It is that the scientists themselves are using different senses. Nobody is actually listening to these signals returning from unexplored regions laden with information. The lab is filled, not with the hollow pinging familiar from submarine war film soundtracks, but with the click and whir of plotters and jocular bouts of repartee. No one now wears headphones and a rapt, faraway look, attentive in ambient hush. For all that modern oceanography relies so much on acoustic techniques, it is machines which do the listening. When I flip a switch on a panel which feeds through a tiny speaker the actual noise of the signals, the American technician Bob sets his face into that expression which in TV shorthand stands for displeasure. "That sound gives me a headache," he says. "It's so goddamn monotonous." I refrain from babbling about hidden subtleties, since they are still there; it is just that they are on an inked printout.

Allowing electronic devices to replace our senses while reducing so much information to visual imagery must have its consequences. Generally speaking, underused faculties tend to atrophy. It has long since become a cliché in the pages of *The Lancet* and the *BMJ* to wonder whether the old-fashioned, prewar family GP with his training in how to watch, to listen, to smell, touch and even taste may have understood more about his patients' health than does his modern counterpart with his reliance on laboratory techniques and diagnostic machinery. Perhaps in dealing with the natural world at an electronic remove scientists in certain disciplines may also risk missing as much as they learn. How many naturalists nowadays have the artist's eye, like the great nineteenth-century scientists who so lovingly sketched their specimens in the field? It is not only sensibility but memory itself which atrophies, since the need for attentive observation is less. The camera takes the place of the eye, the recorder of the ear, the computer of the memory. A laconic finger on a keyboard summons up data, an image. Less need, less time now for Edward Lear's scrupulous parrots or Audubon's American birds, for the hundreds of sketches made aboard the *Challenger* or for anatomical drawings as fine as Jan van Rymsdyck's of the human uterus. Nor is there much call for writing that describes specimens as Philip Gosse described *Cleodora,* a tiny snail known as the sea butterfly which floats in tropical oceans. "A creature of extreme delicacy and beauty. . . . The hinder part is globular and pellucid, and in the dark vividly luminous, presenting a singularly striking appearance as it shines through its perfectly transparent lantern."[2]

[2]Philip Henry Gosse, *A Textbook of Zoology for Schools* (1851), p. 220. Gosse was later criticized by his son Edmund in *Father and Son* for having seen "everything in a lens, nothing

Roger agrees that until well after World War II sonar operators and scientists had spent much of their time with headphones on, listening to the seabed and to anything else whose noises fell within humanly audible wavelengths. They gave their own nicknames to certain familiar sounds, especially those which resisted all identification. One of these became known as the "North Atlantic Boing."

"Now and again something uncanny happens which makes you wonder about what's down there. We've sometimes picked up signals which aren't GLORIA's own echoes but *imitations*. The electronic analysis is quite clear. They're definitely some kind of deliberate response. But what could possibly imitate a sound as complicated as a correlation signal? There are really only two possibilities: either a submarine or whatever creature produced the "North Atlantic Boing" while flirting with sonar operators. Sometimes even our other sonars—the "fish," for example—provoke an almost angry response. We've had the single bleep of the ten kilohertz being answered by a double bleep, or a triple, even a quadruple, as if something's deliberately mocking it.

"You've got to remember that sound travels well in water and we're making a godawful noise down there. We're sending out four different signals powerful enough to bounce off the seabed thirty kilometers away, two of them capable of penetrating hundreds of meters into it. That's a huge amount of sonic energy and we must be absolutely deafening, maybe even lethal, to some animal species. Certainly a whale would find us audible for hundreds of miles, perhaps a thousand.

"So who or what is the author of the 'North Atlantic Boing'? What creature are we enraging or provoking to mimic us? I presume cetaceans. We haven't the first clue what goes on in a whale's mind. We don't even know what's in a *parrot's* mind."

At this point a very sober, older scientist breaks gently in to refute Roger's account of the "Boing." He has the air of someone who has long ago signed the Official Secrets Act and has never regretted it. It makes his explanation oddly dark, vague with the pregnant ellipsis of

---

in the immensity of nature." Yet his descriptions were both lyrical and accurate. That his sense of wonder served to reify an avowed religious purpose (the *Textbook* was published by the Society for Promoting Christian Knowledge) is unimportant. What counts is that he wrote about each organism with an affectionate eye, as though seeing it for the first time: the precise quality which a textbook serves to annul.

someone in the know who cannot tell all he knows. It comes as a reminder that the military has its own interests in oceanography and a good deal of knowledge and technology overlap. There was never any such thing, he says, as the "North Atlantic Boing." The fact is, it was only ever audible on one particular occasion off one particular part of the east coast of the United States. Significantly close, in fact, to one of the major naval bases. What is more, GLORIA uses what used to be a naval signal on the same carrier frequency. Quite possibly it was a submarine communicating with an underwater beacon.

As for Roger's cetacean theory, there are a good few holes in that, too. All cetaceans are far-ranging and one would expect this "Boing" to have cropped up in widely separated areas. Secondly, GLORIA's 6.5 kilohertz is in the five- to ten-kilometer range, and why would any cetacean be interested in anything at that distance? As for the allegedly angry responses to the "fish," Roger will—he is certain—recall a good deal of active porpoise noise in the Bering Sea, especially at ten kilohertz. There was indeed enough energy to break through the correlator return, but it was continuous "active" noise, not a genuine return signal, and it was also associated with biological activity in the water column. It could well have been caused by numbers of animals all fleeing the GLORIA sound at the same moment. Nor should we forget that porpoises can produce all sorts of noise and interference merely by breaking the surface of the water as they swim. Air bubbles are one of the richest sources of noise—look at the swim bladders of hatchet fish—and sudden frothing and foaming plays havoc with signals and can produce the oddest effects. . . .

This is not dissension, of course. The man merely drifts amiably away, leaving behind a feeling of official cold water having been poured, as well as a suspicion that perhaps not everything has been neatly explained after all.

The next day it is discovered that existing charts of this area, which like most maps (other than those of pirate treasure) look completely authoritative, are quite strong. Whole features are either absent or misplaced. We examine the printouts. Entire mountains flicker in and out of existence somewhere down there in the cold darkness. The bathymetry is all haywire. GLORIA knows best. There remains the experience of being present when a portion of the Earth's surface is discovered. This is a rare sensation for the layman in the late twentieth century, and the diminishing opportunities for experiencing it must belong almost exclusively to potholers and cavers, apart from oceanographers themselves. At this moment, except for the handful of people present in the *Farnella*'s laboratory, I know more of the range of hills we are traversing than anyone else in the world, be they professors, explorers, Nobel laureates, fellows of distinguished societies or captains of industry. In fact, I know more of this piece of America than the president himself, and I am not even a US citizen.

This schoolboy superiority is too brief and dubious to be at all satisfying. Moreover, it is imbued with a certain sadness. One more thing has fallen under *Homo*'s rapacious gaze, and as always the knowledge is not neutral. By its very nature this project makes one constantly aware of the question of ownership. Effectively, anything we find down there belongs to the United States which, by annexing its tranche of the seabed, is actually adding a prodigious 2.9 billion acres to the 2.3 billion acres of dry land it already commands: rather more than the same again. Quite apart from any military consequences, the economic aspects are plain. This makes the fixing of all EEZ boundaries a matter of great importance, both scientifically and bureaucratically. As soon as one starts to consider it in any detail, obvious difficulties present themselves. On this cruise we are supposed to be mapping the EEZ around Johnston Atoll. Are the 200 nautical miles to be measured from the island's coastline or from its center? There again, if they are measured from the coastline does the boundary have to describe a neat circle or should it faithfully follow every little promontory and indentation so that on a chart Johnston Island will eventually appear as a small blob in the middle of a dotted outline of its own hugely magnified ghost?

According to Roger everything depends on the size of the island. International law stipulates how big an island has to be before the EEZ may follow its coastline. There again, a mass of guidelines define a coastline and how to treat it. If the mouth of a bay is less than a certain width it must be considered as if a straight line ran across it. Such niceties are far from being purely academic. According to Colonel Qaddafi the Gulf of Sirte is entirely within Libya's territorial waters whereas the US Navy, which is constantly on patrol there, claims the Gulf is international.

"It usen't to be like this," Roger says. "When I first started in oceanography—and this is only about ten or fifteen years ago—you could still go where you liked to do your scientific cruises, just as they did in the last century. Nobody ever objected if a British survey vessel chose to do some sounding or coring forty miles offshore. Nowadays if we want to do fieldwork inside somebody's EEZ we have to apply to the Foreign Office, can you believe, to get permission on our behalf and like as not we'll have to agree to take appropriate foreign observers on board. It's a ridiculous hoo-ha.

"It's all about wealth, of course. Round here there are some huge fields of nodules that are potentially rich pickings. It's up to us to locate them precisely. But here, for example"—he points to a chart taped to the bulkhead—"southeast of where we are now, that's the Clarion-Clipperton Fracture Zone, famously rich in nodules. As you can see, it's in the middle of the Pacific. It doesn't fall into anybody's EEZ. Yet in the early eighties President Reagan declared it to be an area in which the US would regulate all mineral mining licenses. Incredible, when you stop to think about it. It's not even US territory. Reagan's argument was that there was a sort of power vacuum there and that *someone* had to supervise things otherwise there'd be all kinds of skulduggery. Pure altruism, you aren't thinking. So the whole thing has now been pretty much carved up by about four consortiums. No skulduggery there, naturally. Now, the question we're asking is, supposing somebody starts mining manganese nodules in an area like that and you come along wanting to do your bit of oceanography? Neither they nor you have an obvious legal claim. It's international waters. But I somehow can't see a deep-sea mining outfit sitting idly by while a fully equipped research vessel heaves to a mile away to do some close investigating of the field."

One afternoon I spend time in the darkroom where Roger is making prints of the laser printer films of GLORIA scans. It is drudgery, but needs constant vigilance. The contrast between the high technology outside in the lab and the low in the darkroom is marked indeed. The ancient printer is suspended on a wooden tray by four cords from the ceiling. The scale of the prints changes fractionally all the time and focusing is critical. The reason is the fundamental cartographer's problem of how to represent a curved surface on a flat one. On the variant of the Mercator's projection the Survey is using, the latitude lines grow farther apart the farther north one goes. At the levels of accuracy required, the *Farnella*'s slightest northward drift can make a difference.

"It's not for fun, all this," reminds Roger's voice in the darkness.

"It's about potential megabucks. We're likely to find nodules here-abouts. If so, somebody else will come along on a sampling trip to see what the quality's like and they'll want to be able to arrive at the precise point and drop a sonar buoy. And they'll be doing it from our GLORIA scans. I know it's a bore, fiddling about with the focus and scale and stuff, but I do quite like that it's a tangible effect of ordinary physics. . . . Hey, you know the Forth Bridge? Well, its two towers are absolutely plumb vertical, but their tops are a centimeter farther apart than their bases. Curvature of the Earth. I love that, don't you?"

The enthusiasm of his voice in this stuffy cubicle full of the reek of developer and fixer is contagious and banishes the monotony of the task. He is apologetic in case his recently completed doctoral thesis sounds dull. "Actually, it's on acoustics and sedimentation. . . . Basically, why is it that GLORIA's signal comes back at all, rather than simply scatter-ing away all over the seabed?" Although still young, Roger is a veteran of field trips all over the world, from Tahiti to Alaska. One of the surprises of modern oceanography has been the discovery of undersea fans and river systems which, as mapped by GLORIA, bear a great resemblance to aerial and satellite images of river deltas and similar land features. This, too, has a commercial aspect. "Take gold. A glacier shield like the one north of Alaska liberates gold from the Earth's crust as it grinds away. This, in the sediment, eventually flows into an undersea river system. Over millions of years that's a lot of gold. But at the same time the crustal movement shifts these ancient deposits about while the actual sources don't necessarily change position. So an old auriferous sediment may today lie deep beneath a shallow modern fan system containing practically no gold at all.

"Now, if you've been looking at GLORIA and seismic scans for years, and happen also to have a geological nose, you can make some pretty shrewd guesses as to where there might be an awful lot of gold. Certainly where it'd be worth sinking a borehole."

"Roger the Klondiker. You're an adventurer at heart."

"I'm sure as hell not staying in academic geology all my life. I want lots more fieldwork, the more the better. I'm thinking of joining this sailing clipper to investigate a fjord system in South America. Lovely stuff: a proper sailing ship. I want to see things."

He has already seen things, including Mount Paramount in Alaska—minus the halo of stars—which any moviegoer would recognize. I remark that this is not the first time he has sounded bored with his present niche in geophysics.

"Well, of course, for plenty of geologists, including some aboard this ship, I may say, it's just a job. They could be doing anything, really. They mostly do it perfectly adequately but with no particular enthusiasm, so they bring no particular imagination to bear."

Nobody ever seems to go up on deck just to look at the sea. They go up to inspect gear or sunbathe obsessively with a dogeared bestseller from the ship's library, but on the whole are not to be found staring into the scud and dazzle and maybe wondering about the strangeness of being in a floating speck suspended miles above mountains, like the silver dot of an airliner over the Alps. Nor, with a few exceptions, do they seem aware of sailing across a wasteland which, even five years ago, would have teemed with dolphin and porpoise and now stretches to the horizon unbroken by anything other than the occasional flying fish. To see the world's greatest ocean suddenly empty within a few years is to be filled with a foreboding which cannot be dispelled. Something is happening below us which geophysical oceanography, its electronic gaze fixed firmly on the lumps and hollows of the seabed, is missing. Yet it needs only to come up on deck and look at the empty waves, the nearly birdless sky.

If the scientists aboard *Farnella* split, unsurprisingly, into those with imagination and those with less, odder fractures become apparent. On a tour of the ship on the first day Roger had showed me the bow thruster room. This led down to a horrible cubbyhole containing cramped catwalks above bilges and the hydraulic columns of an electric motor and auxiliary propeller which could be lowered through the hull for fine maneuvering when core samples were being drilled from the seabed. It was into this cubbyhole that, last 4 July, a British oceanographer had lured some American colleagues and closed the hatch on them, lowering gruel through a hole and placing a sizzling steak on a grid just beyond reach. After two hours he alerted a hapless seaman to odd noises coming from the bow thruster room and made himself scarce as the released scientists roared through the ship. The incident of the bow thruster room had entered the unofficial annals of oceanography even as, at the other end of *Farnella*, GLORIA was patiently mapping an unknown section of the Planet's surface. Pranks and discovery.

"They take themselves so *seriously,* the Yanks," Roger observes in what he thinks of as the confidentiality of the darkroom. This Independence Day caper was recounted as a lark, not as savage. Still less was it an act which had quite precisely emphasized dependence. Aboard this ship the British are in numerical superiority and sure of themselves. The knockabout humor is such that a female scientist is required to become one of the lads, to the extent that when a woman geologist refers to her fiancé in San Diego I am momentarily as surprised as I would be if the bos'n himself had alluded to a boyfriend in Dutch Harbor: surprised not by the fact, but by the evidence of any adult life elsewhere. One evening a dozen of us are watching a ho-hum video film about a hardboiled American cop who goes to Japan to track down the Yakuza gang member who offed his buddy. This provokes one of the British scientists into giving a commentary in "Japanese" (high-pitched exclamations, aspirated "h" sounds, *Carry On Samurai* syntax and all). The strange thing is that Elly, a Japanese-American geologist, is sitting three feet away. She elects gracefully to ignore the whole thing. From time to time throughout the trip this most amiable man's hysterical Nipponese, presumably quite unconscious, can be heard now in the lab, now in the dining room (brussels sprouts apparently being reminiscent enough of bean sprouts to set him off). Elly betrays no sign of offense except for now and then shooting him a weary glance. Maybe oceanographers prefer their bonhomie ruthless. This is not the first occasion that she and several others, including this fellow, have sailed together; conceivably she has built up the necessary tolerance.

Elsewhere on *Farnella* are reminders that intellectual lives are being lived and thoughts thought. I am tackled at breakfast by Stuart, an electronics technician from Burbank who over scrambled eggs wants my opinion on Sacheverell Sitwell, Proust and Herman Wouk. He is also a musician. "Dinu Lipatti. God, nobody will ever play the B-flat Partita like Lipatti. Not *ever*. And what about Earl Wild? Fantastic technique. Do you know his recording of Liszt's *Don Juan* fantasy? Isn't it great? And there's an incredible track on the same disc, another Liszt fantasy. It's on a Meyerbeer opera. . . . No, it's gone. I get really mad when I can't remember . . . *Robert le Diable*, that's it." Stuart is in his mid-fifties, looks twenty years younger. He has a wife and houses in California. He is no stranger to life at sea, having spent several years in the US Navy. "But I've got to get out of this business. I always fall foul of hierarchies. I never can get the politics right. I can't work out why people so much my junior always seem to wind up calling the shots. Maybe I'll start a business repairing all those videos people just junk because there's nobody to fix them. Easy money. The rest of the time I can read and play music."

Strange shipboard rituals, too. The sea has always presented itself as a collection of symptoms while denying itself the status of a disease. Every evening before supper ("tea," between five and six P.M.) the "Oily-Boily Bar" is convened. The name refers to oily boilersuits and means that anyone is free to come and drink in the democratic surroundings of an engineer's cabin. That is, a scientific engineer, not a crew member. ("Them and Us?" repeats one of the cooks thoughtfully. It seems to strike him as a pithy, even novel, concept. "Yes. Them and Us. I reckon you could say that.") The demarcation is not at all snobbish. Many of both crew and scientists are veteran shipmates of several cruises and greet each other amicably when they meet in passage and companionway. I simply record that I never saw a crewman at an Oily-Boily Bar session.

Another semiritual is "Hump Day," the cruise's halfway point "from where it's all downhill," as somebody says. I had always associated the expression with old hands of the Far North—whalers, Yukon trappers, Arctic explorers—as meaning the moment when the endless winter reaches its shortest day and the light at last begins to lengthen. Hints of desperation, too, in "Gangplank Trials," when in the last week of a long cruise the gangway is reportedly laid out flat on deck while scientists in full shore gear run along it carrying suitcases. "It's a rehearsal, isn't it? Helps the spirits. Tells you it's not long now." The echo is of the "Channel Fever" which returning British sailors used to contract on first glimpsing the south coast, knowing they were still a couple of days or more out of Liverpool or Glasgow.

Belatedly it becomes clear. Probably none of the younger scientists on *Farnella* has been to a boarding school, in the military, or even to prison, and they lack a way of measuring time internally. They are mainly family people, not seafarers at all. Their business happens to be oceanography rather than metallurgy. Somewhere in Surrey and San Diego are office doors with their names on them, lockers and lab smocks belonging to them, named spaces for their cars outside, a house within easy commuting distance. They are not gypsies or nomads.

~

The days pass, the ship goes back and forth along its lines, the shaded portion of the circle grows. No disasters, only the single hitch of the weather. The wind is stiff, the waves high, as if we were perpetually on the edge of a storm system 500 miles away. The seas hit us at an angle, nudging the bows off course. There are worries about possible damage to GLORIA's cable, which thrums like a steel bar over the stern, slackening in troughs and then tautening with a snap. The scientists confer with the captain and agree to knock a knot off the speed. This means computer work so the scans are not distorted. A new seamount is discovered whose foothills were first spotted on the previous leg while traveling in the opposite direction.

"I don't know what the hell that is," says Mike, pointing at a gravelly-looking portion of a gravelly-looking picture. Doctors poring over an X-ray.

"I reckon it's a nodule field," says Roger. "I've seen something just like it elsewhere. Those are really sheer cliffs [rock faces of 3,000 feet which the world's mountaineering community would race to get their pitons into were they on dry land], and there's less sedimentation here so we're in the lee of the local current system. That means *this* stuff"—a chewed ballpoint taps the picture—"is boulders. Detritus that's fallen over the edge or sheared from the face. Doesn't look like any new activity here, so these'll be nodules, half buried by the debris and extending out to—wow, it may be off this leg too, so it's a big field. Anyway, bet you anything that's what it is. I feel it in my bones."

"And can Roger's famous bones be wrong?" murmurs Mike.

The tone of these conversations, the friendly offhand speculativeness, betrays like nothing else the distance traveled over the last hundred years by the Earth sciences, and particularly geology. The casual references to old and recent rock formations are maybe not novel, but the confidence belongs to an era which has an unassailable timescale fixed in its collective scientific head. The seabed we are mapping is not immeasurably ancient. The area we are in, not far from the Hawaiian chain, is volcanically active and the seabed comparatively mobile. Johnston Atoll, for instance, is a ruin of its former self, far smaller and sinking steadily. (Its reefs show it is sinking just slowly enough for coral polyps to keep pace.) Even on dry land the most monolithic features often turn out to be quite recent affairs. Not far below the summit of Mount Everest are chunks of oceanic crust with fossil sea shells embedded in them, indicating that the entire Himalayas were a stretch of seabed until India banged into Asia and the impact pushed the rocks five miles straight up into the air. That was a mere forty million years ago, in geological terms practically the other day.

Aboard *Farnella* there is easy talk of subduction zones; any disputes arise over quite specific technical theories associated with them. Subduction zones (where the edge of one tectonic plate dives down beneath the edge of its neighbor) were not suspected until plate tectonic theory proposed them, and tectonic theory itself was a direct descendant of the theory of continental drift which the German Alfred Wegener had brilliantly proposed in 1915. Poor Wegener! In 1928 a panel of fourteen geologists was convened to vote on his theory and only five were fully in support. Two years later he vanished in a blizzard in the middle of Greenland and his theory suffered a similar fate until the 1950s. I myself can remember a geography lesson in early 1955 when the teacher, in response to a question from a boy who must either have been very studious or else a troublemaker, suddenly shouted: "Continental drift is *bunk*! It is *loathsome* bunk!"[3]

And so, as the geologists speculate and tap their bitten ballpoints on computer printouts, I have to remind myself that not so long ago their conversation would have been pure heresy and have evoked tirades of condemnation, not only from other scientists but also from fully robed bishops. A bare hundred years before I was born, William Buckland (who was both a geologist and the dean of Westminster) wrote his *Bridgewater Treatise,* in which he categorically affirmed that Noah's Flood accounted for sedimentary rocks and fossils. To watch the endless scrolls of paper emerge jerkily from printers and plotters is to feel that this is not so much mapping a new world as burying an old, which in turn makes one wonder what fresh heresies the future holds.

[3]Yet there is still dissent. Dr. Glebb Udintsev at the Moscow Institute of the Physics of the Earth consistently refuses to believe in plate tectonic theory, specifically rejecting the idea of subduction. His view is that the Earth is slowly expanding. This would indeed explain many tectonic phenomena, though much seismic data would have to be ignored or tendentiously interpreted. Unfortunately, it is not entirely easy to disprove the expanding Earth theory since any rate of change would be infinitesimally slow, too much so to be revealed by bouncing and timing radar signals off the Moon, for example.

Meanwhile, the question of whether or not we have discovered a hitherto uncharted field of manganese nodules looks like having to remain undecided until a later cruise. The weather is now bad enough to make us abandon our course and begin a new set of legs on a different tack.

"Not so nice for us, though. It'll mean rolling instead of pitching. But kinder to the equipment."

"What does the forecast say?"

"We *are* the weather forecast, that's the trouble. The *Farnella*'s an official weather-reporting ship when she's on station, so the forecasts out here are simply based on our own data from yesterday."

I like the idea that we are helping invent the world's weather but can see it is unhelpful having no higher authority on which to rely. Still, the measured tones of a radio weatherman repeating one's own reports might well endow them with an official quality which would suddenly make them credible as predictions. Up on deck it is exhilarating as the ship buffets into the wind, shouldering clouds of water back over her superstructure. Another day denied the sunbathers, their towels unspread, their copies (courtesy of the Marine Society) of Tom Clancy, Len Deighton, Clive Cussler, Cruz Smith, unopened. The library contains dozens of these more or less identical Cold War what-ifs: unmemorabilia which after the last few years have suddenly come to seem like fossils, requiring of their readers a streak of the antiquarian, even of the geologist. By contrast, nothing could be more present than this ocean leaping past and over. It raises its crest to the horizon, empty and brisk, at once a locus of sublime vacancy and massive energy. When the tropic sun blazes through cracks in the scudding clouds a brilliant ultramarine floods back into the water. Simultaneously, its tusks of foam gleam white while peaks and crevasses wring colors from the constant motion. At these hot, radiant moments the sea is transformed into prodigious and arcane machinery, liquid clockwork glinting with moving parts. In the intermittent bursts of light it looks like what it is: the Planet's gearbox mediating and transmitting the motive power of the sun.

In early afternoon the noise of lighter engines detaches itself from the background roar of the ship's funnel. An unmarked gray P3 Orion flies past very low off the starboard bow, its two inboard propellers feathered. Ahead, it makes a wide sweep, banking so its wings flash against the storm clouds like a fulmar's, then returns for another pass. Coming from in front its crew will easily be able to read the great blue logo painted below *Farnella*'s bridge: a circle containing an anchor and the legend "US Geological Survey." This identification will be superfluous, since the aircraft is one of the US Navy's long-range antisubmarine patrols and will automatically have dropped sonar buoys to learn what we are up to. They will already have heard our cacophony of pulses and identified GLORIA's correlation signal. A hand is raised behind a cockpit window as the plane drones by. It climbs away to the west trailing dark exhaust and is soon lost in the cloudbase. The ocean feels momentarily emptier for its departure.

There is a real oddness in all this watching and listening. Held firmly in the US Navy's electronic gaze as we are, the *Farnella* is herself sizzling with codes which enable us to listen to the seabed and define it, while undoubtedly provoking the attention and maybe even aggression of untold creatures below. Nobody is listening out for them at the end of this bizarre acoustic chain, however, and they will only be heard if the wavelengths of their signals happen to coincide or interfere with those of the scientists' finely tuned sensors. Certainly no human ears are on the alert for them. In the meantime we are also listening to ourselves via our own weather reports. The ocean which surrounds and supports us manages not to impinge on our senses except to induce queasiness in one or two people.

"I'll never go on another geophysical cruise, I swear it," Pattiann says next morning. "They're so *boring*. A sampling cruise is much more interesting. The dredge brings up lovely great chunks of stuff. At least it's something to paw over and get your hands mucky on. All these computers—that's not my idea of geology."

She assumes I am as bored as she, but I deny this.

"Still, next time I'd go on one of our sampling trips," she recommends. "Or a JOIDES drilling and coring cruise where you get to see real stuff from below the bottom. Bits of the living Planet, you know. This electronic surveying's too hands-off for people like us."

Pattiann, who must be somewhere in her forties and hence among the older of the scientists aboard, may be making a rhetorical gesture implying a companionable solidarity, or else the more radical point that scientists' differing views of the technology at their disposal might have much to do with their age. Pattiann's lineage is from the nineteenth-century oceanographers, who were excited by touching and smelling and tasting whatever they studied. In fact, a certain skepticism about modern high technology, not least its prohibitive cost, is detectable in all but the youngest scientists at one time or another. Obviously there is no cheap way to map a seabed accurately from a distance of five kilometers; yet an elusive feeling of there being a discrepancy somewhere lies on the printouts like a glaze. It is as though discovering what there is beneath the sea, learning about places on Earth which no one has ever seen, ought to be momentous, even personal. Yet it is largely automatic, done even as we sleep. Machines make the signals, listen for their return, gather navigational data, juggle it around in digitized fashion, convert the digits to a visual analog and produce an image for a geologist to tap his teeth over. All well and good; but to requote Roger's maxim: In science, as elsewhere, you can't get something for nothing. At some point there is always a debt, a deficit, a loss. A law suggests itself, to the effect that the way any data are read is a function of the way they have been gathered.

~

This cruise is unusually short, a mere fortnight at sea. It has been timed so people will be able to fly home for Christmas. The mosaics of the GLORIA and air-gun prints grow on the chart table. The sweet, hot-blanket smell of popcorn drifts from the other end of the lab, where a machine is anchored by Velcro strips near the swaying gravimeter. Katie stands her watch in the bay which houses the electronics for the two midships sonars. Or rather, she mostly sits. Only occasionally does she need to get up and jot some figures with a felt-tip pen on the printouts inching their way from the plotters. She clicks a switch or two, changes a stylus, checks the course, goes back to her book.

It is pleasing to think that most of the ocean bed will never be seen directly by mortal eye. In comparison to that vast area the ground which might be covered (at such prodigious expense) by manned submersibles is virtually nil. One might compare it to traveling across Asia by oil-lit hansom cab with the conditions of a Dickensian fog outside, and then claiming to have seen the world. At the same time, sitting in *Farnella*'s lab and looking at the banks of instruments, computers, screens, laser printers, plotters, popcorn poppers and their umbilical harness of cables, all of it sailing steadily along a knife-edge course five kilometers above the Late Cretaceous seabed, one cannot help reflecting on *Homo*'s fierce if limited intelligence. The remorseless taxonomy of a century ago, using primitive but ingenious measuring devices as well as guns and nets and killing bottles, seems far distant now; yet even that distance is as nothing compared with the exponentially opening gap which separates *Homo* from the rest of Earth's organisms.

Katie is doing her watch early because 18 December is her birthday. Since she is known to be fond of rabbits there are celebrations tonight with rabbits as the theme. We have already signed a joint card and are encouraged to cut loose with the fancy dress. "Rabbits, you guys. Anything goes. Raid the hold." Not for the first time a deep current of hysteria is apparent and in these final three days of the cruise becomes a good deal closer to the surface. Katie's colleague Sue is for this trip the scientist nominally in charge of representing the US government's inter-ests. Despite the protestations of powersharing, democracy and coopera-tion, somehow or other it is Sue's judgments alone which have been prevailing this last week. She has a career and a CV to think of. A failure to complete a geophysical survey for which she was responsible will be remembered, if not held against her. Unknown to the rest of us she has taken a chance. Despite the weather and *Farnella*'s reduced speed, Sue has insisted we complete our allotted task even if it means cutting things fine.

At tea next evening the news reaches us from the bridge. There is now no way the ship can arrive back in Honolulu on the Friday morning as planned. Indeed, if the weather gets any worse we could be anything up to twenty-four hours late. Suddenly, there is a possibility that everyone may miss their flights home. Christmas Day is on the following Tuesday and the flights had anyway been hard enough to book. Over peaches and custard and the remains of Katie's birthday cake all pretense of the cool, scientific approach evaporates. Hard words are spoken about stubborn and cocksure youngsters whose inexperience jeopardizes decent folks' family lives. The scientist who so recently stood her GLORIA watch slightly the worse for drink and wearing a long pair of cardboard rabbit ears puts down her spoonful of custard and stares at her plate, cheeks crimson. It is pure boarding school. Then the US government's geolo-gist bursts into tears and runs from the room.

"Huh," says a heartless Brit, "we'd all like to be able to do that, wouldn't we? Do *we* feel any better about missing our Christmas?" And looks round rhetorically for assent. A deputation of girls plods off to the lab to comfort Sue, who has taken refuge among the charts and plotters of her disgrace. "Right, then. I'm off to the bridge to kick ass. Anyone coming?"

Presumably each expedition becomes characterized by its own catch-phrase. This cruise has acquired two, one of them written up on the lab's steel bulkhead in bar magnets, "Yee Haa!": a cowboy's yell which flew one night out of the drunken Oily-Boily Bar. The other phrase is "Kickin' ass," which has recurred on surely every one of the video movies we have sat through this last fortnight—movies about cops and cops and cops, winsome black cops and shitty white ones, marine sergeants and top gunners—so that suddenly the whole of American culture seems embodied in a catastrophic anger. It is with these furious heroes we are supposed to identify, these men with their scratched biceps and bared chests and 400-word vocabularies; so that a mild geologist from Godalming, put out because he may not make Honolulu on time and hence his flight back to England for Christmas with the family, says "I'm off to the bridge to kick ass."

During the next day, though, things look up. With all the equipment safely back on board the *Farnella* is able to pick up speed. The chief engineer, himself due in New Zealand for Christmas, is coaxing every last revolution out of the engines. ("All right for him," says his junior. "He's retiring after this trip. I'm the poor bugger's got to put in new piston rings over the holiday.")

In the event we dock at six P.M. on Friday after all and nobody misses their flight. Sue is back to being one of the boys. GLORIA is back in its cradle. The rolls of printout, the reels of computer tape which are the only tangible evidence of the invisible seabed we have been crisscrossing for the last two weeks are safely packed up. It has all been a great success, is the verdict. No equipment lost, nobody swept overboard, unlike the luckless oceanographer who had disappeared recently one stormy night in the Bristol Channel. In fact, a cushy number all round. Even the sea has a satisfied look to it as it mulls around the pilings of Honolulu harbor. It has so simply kept all the secrets it had which were worth keeping.

Cabs arrive on the quay to take the scientists on a last-minute shopping spree before their flights next morning.

"Off to Hilo Hattie's to get a really *crucial* pair of shorts," is Roger's valediction. They vanish in a cloud of exhaust. Stuart appears at the rail next to me, slightly mournful in shore kit.

"Do you know Rosalyn Tureck's performance of the Forty-eight? What do you think? Total contrast with Gould's reading, I guess. Some of his tempos seem downright crazy but God, I remember when his first Goldberg came out in the fifties. We'd never heard Bach playing like it. Nobody had. The energy! Everyone thought the Goldberg was dead, academic, cerebral stuff, you know? But it wasn't. It was *alive.*"

He stares down at the crack of water between wharf and hull. His voice is more animated than at any time in the last fortnight.

2. "NOTHING IS MORE TEDIOUS THAN A LANDSCAPE WITHOUT NAMES"

Immediately to the north of Hawaii, scattered across the Murray Fracture Zone, lie the Musicians Seamounts. They stretch for maybe 200 miles, from Strauss in the north to Mendelssohn in the south. There is a Bach Ridge and a Beethoven Ridge. There is also Mozart, a considerable mountain rising from the abyssal plain five kilometers below to within 900 meters of the surface. Mount Mozart, while a fairly minor affair by suboceanic standards, is therefore slightly taller than Mount Fuji, although of nowhere near such classic proportions.

The presence of this random clutch of composers *engloutis* in the middle of Pacific wastes is a reminder of how much of the physical world belongs in its taxonomy, description and name to the Western nations. It is also a reminder that in a sense things do not exist until they are named. Before that, everything partakes of a state of undifferentiated chaos which is never a neutral matter to human beings but carries a degree of menace. To name something is to take control of it. It could be argued that the Old Testament story of Genesis was less a matter of creation than of naming, of God taking control of chaos. Whereas before, the pre-Universe consisted of a kind of primordial babble, God-grammarian sorted out its constituent parts and uttered some solid nouns—dualities, mainly: crude oppositions such as light/dark, heaven/earth, sea/land. How he had entertained himself before this basic act of intelligence is open to speculation, but if he was anything like the humans he created (and according to Scripture he was) he was bored, repelled and finally menaced by a universe which was still a state rather than an infinite collection of objects. Ever since, *Homo* has felt the same and travelers have gone about the globe as adventurers, conquerors, sightseers, nomads and scientists, naming its parts and often bestowing on them their own proper names as well as those of their friends and sponsors. In his novella *Colomba* (1840), Prosper Mérimée's English heroine, Lydia Nevil, takes pleasure in learning the names of places on the Corsican coast as she passes in a schooner, for "nothing is more tedious than a landscape without names." Many a sea captain found his spirits insupportably lowered by a coast such as that of Africa, when whole days might go by without sight of a single named feature. It would presumably have made little difference knowing the local tribespeople had their own names for the hills and capes and rivers. Being illiterate, they would have been ineligible to bestow valid names because unable to write them on a map. Only cartography can remove names from merely local usage and bring places into international being.

The desire to tame a threatening landscape by subjecting it to the control of language can be seen in the old Greek name for the notoriously treacherous Black Sea: the Euxine, or hospitable. An extension of this may result in the temporary renaming of already well-known places. In World War I when British troops were mired into the static and murderous wastelands of trench warfare, micromaps were devised for the tiny localities which bounded their lives. London place names were wistfully bestowed on slivers of Belgian and French farmland. What a year or two earlier had been "Quineau's acre" or "Drowned-cow bottom" were now Haymarket and Leicester Square. This yearning domestication of threatening foreign places is a common enough trope in wartime ("Hamburger Hill") and came equally naturally to Pincher

Martin, William Golding's wrecked sailor. Almost his first act on being able physically to patrol the Rockall-like Atlantic islet on which he was washed up was to give its features familiar names like Prospect Cliff, High Street and Piccadilly. This was in recognition that, unnamed, the place of his marooning would have remained inimical to him as well as invisible to rescuers, being quite literally off the map.

A Mozart Seamount does, however, seem particularly arbitrary in the subtropical latitudes around Hawaii. Odder still, it is equally close to Gluck and Puccini Seamounts, just as Haydn is to Mussorgsky and Beethoven Ridges. Clearly it is useless to look for any correlation between the physical proximity of these seabed features and the chronology of their namesakes. Somebody must have thought "We've done poets, now let's do composers," much as local councils name the roads of new housing estates. It is only since the invention of a technology powerful enough to map the deep seabed that the finding of names has become a pressing issue. By the early years of this century most of the Planet's territorial features had been mapped and named, with the exception of the remotest hinterlands like Antarctica and the Amazon jungle. Sidescanning sonar is now revealing ever more details which for geologists, if for nobody else, need to be identifiable by name. As far as the military is concerned the situation remains equivocal. Strategic seabeds like that beneath the Arctic icecap have been extensively mapped by NATO and Warsaw Pact submarines, but their charts remain classified. There are projects for civil mapping and geological surveys of North Polar waters, but they remain projects until somebody donates a nuclear submarine to an oceanographic institute.

In order to cope with the need for new names on new charts there are various regulatory bodies which amount to a more or less official international committee on names. There is, for example, BGN/ACUF: the US Board on Geographic Names, Advisory Committee on Undersea Features. There is also the Monaco-based GEBCO: General Bathymetric Chart of the Oceans, an organization founded at the beginning of the century. These bureaucracies are constantly turning out documents, indexes, guidelines, lists of eligible names and the like. Very occasionally a lone human voice cuts through it all, like Robert L. Fisher's in his "Proposal for Modesty." In this he inveighs against

parvenu scientists who offhandedly baptize a deep-sea . . . feature that may have been known and well-explored—even if possibly unnamed—earlier, or even one bearing a long established name in another language. . . . Some . . . apparently know so little about historical courtesy, significant commemoration, or even good taste that the seafloor is becoming littered, and the literature of marine geology and geophysics cluttered, with personal, in-group, self-aggrandizing, back-scratching, trite unimaginative ("14°N Fracture Zone") names or ugly acronyms ("GOFAR Fracture Zone").[4]

The Musicians Seamounts are an example of bureaucratically approved naming. It was likewise decreed that a group of submarine features off the southwest tip of Eire should be named after Tolkien characters, which explains the Gollum Channel. The bureaucrats do not have it all their own way, however. Now and then the working names which pioneering geologists assign their discoveries stick, in all their whimsicality. A few years ago Quentin Huggett and his IOS colleagues were mapping some seabed fields of manganese nodules with GLORIA when they found a series of hills which they needed to be able to identify as they worked. One became Nod Hill, a second (felicitously named on Christmas Day), Yule. A third hill became Mango while the fourth— unfortunately never discovered—would inevitably have been Knees. Nod, Yule and Mango Hills remain to this day and probably always will, long after they have been stripped of the asset which gave them their name, like the Gold and Ivory Coasts.

A more famous and no less whimsical example is of an area of Atlantic seabed to the west of Spain which celebrates British biscuits. This centers on the Peake Deep, modestly named after himself by the ship's captain who discovered it. A later expedition from Cambridge found a long, shallow depression in the same area which they loyally named King's Trough. Then they discovered a second deep near Peake Deep and called it Freane [*sic*] Deep. Further surveying disclosed two ridges between these features which became respectively Huntley and Palmer Ridges. Finally, the trip was completed with the identification of Crumb Seamount.

More recently in the Aleutians, while mapping the 200-mile EEZ around Alaska, GLORIA at the end of one of its turns revealed an unknown volcano beneath Soviet waters. Quentin Huggett, interested

[4]Robert L. Fisher, writing from the Scripps Institution of Oceanography in *Geology* (June 1987).

in pre-Soviet Russian anarchist movements, reported its existence to the Soviet Academy of Sciences with the customary apology for unintentionally having "spied" into Soviet waters and suggested it should be called Kropotkin Seamount in honor of Peter Kropotkin, the celebrated geologist and anarchist. It so happened that Kropotkin's nephew, himself a geologist, was on the panel of Academicians considering the suggestion and reportedly the proposal raised a laugh in the relaxed climate of *perestroika*. A certain edge to the laughter might have come from the knowledge that Peter Kropotkin had been distantly related to the Romanovs and his nephew is considered by some today to be the Soviet citizen nearest in succession to the Russian throne. In this particular case, and beneath the international exchange of jocularities, curious games are perhaps being played. For while the implication of the story—heard from the Western side—is that scientists with superior technology could tell a country things it does not know about its own territory and would happily do so for a price, the story from the Russian side might be quite different. With military security it is never quite certain what is known. The Bering Strait must after all be one of the areas most familiar to Cold War submariners, and it would seem likely that this seamount was already known to the Russians, who might have declined to submit it themselves for international naming in order to disguise the extent of their own knowledge.

3. ZETETICS

This chapter has constantly invoked the notion of a globe, an oblate spheroid, to represent the Planet. This is reasonable, given that not only do Newtonian physics and mechanics define it as such but it looks like one when viewed from space. There has long been—and still exists—a pseudoscience called zetetics which maintains that the Earth is in fact flat. Its adherents are popularly known as "Flat-Earthers," a term of disparagement which connotes either stupidity or else wishful, head-in-the-sand archaism. Nor is this scorn unreasonable, given the crackpot tone in which their case is usually advanced.

When zetetics most earnestly offers its evidence, the classical procedure is for it to cite a list of mathematical and other "conundrums" which might be taken as casting doubt on Copernican theory. It first proposes a model of the Earth which is a vast disc, an irregular plane of unspecified thickness and circumference at the center of the universe, above whose surface the sun and stars circle on concentric paths. Its circumference is indeterminable because the edge of the known Earth is surrounded by a barrier of ice (which others might call the Arctic and Antarctic) beyond which "the natural world is lost to human perception. How far the ice extends; how it terminates; and what exists beyond it, are questions to which no present human experience can reply." The words are those of "Parallax," the second edition of whose *Zetetic Astronomy* was published in London in 1873. There is reason for thinking he might also be the S. B. Rowbotham who published a book of the same title in 1849. This earlier date has some significance because it falls in a period of great interest and debate about the age of the Earth (see p. 195) and it is not at all surprising to find that "Parallax" is a firm believer in 4004 B.C. as the date of Creation.

He begins by describing several experiments with flags, poles and ships to prove that the surface of water—and therefore also the sea—is not convex, and in due course reaches his cannon test. His argument is that if Earth were a rotating sphere a cannonball fired vertically into the air could not possibly fall back on top of the cannon. There is an engaging, schoolboy quality to this idea, like that of a child who imagines himself falling in a broken lift but able at the last moment to save himself (unlike everyone else in the lift) by craftily giving a little jump just before it crashes at the bottom of the shaft. At any rate, the conundrum of the cannonball exercised the minds of some quite elderly schoolboys during the Crimean War, and on 20 December 1857 the prime minister, Lord Palmerston, wrote to his secretary for war, Lord Panmure, to clear up a few niggling points about British gunnery. He was assuming a cannonball fired in the air would not follow exactly the rotation of the Earth's surface but would to some extent be left behind. That being so, he wondered if the secretary for war realized that the tactics of modern warfare ought perhaps to be altered to take into account the obvious fact that the range of guns must vary according to the direction they were pointed in. Clearly, if they were fired eastward in the direction of the globe's rotation the balls would "fly less far upon the Earth's surface than a ball fired due west."[5]

"Parallax" reexamined the accounts of the voyages of oceanographers and explorers like Maury and Sir John Ross to show where their naviga-

---

[5]"Parallax," *op. cit.*

tion had been at fault. It was hardly surprising that a man like Ross, even though of the highest personal integrity, had been deluded into thinking he had spent four years completing a circumnavigation of the globe when all he had done was sail 69,000 miles around the inside of the Great Ice Barrier. There was an urgent need to revise the whole science of navigation, particularly knowledge of the tides, sunrise and sunset, the seasons and the laws of perspective. Some of the writer's own conclusions about the heavenly bodies were indeed radical. The sun is "considerably less than 700 statute miles above the earth," and "all the visible luminaries in the firmament are contained within a vertical distance of 1,000 statute miles." Moreover, the Moon is transparent. "We are often able to see through the dark side of the moon's body the light on the other side."

This is most inventive, and "Parallax" had many disciples from that day on, some of whom cribbed his examples. In 1940 a certain E. L. Venter of Bloemfontein published *100 Proofs That Earth Is Not A "Globe"* and also considered the cannon test, concluding "the ball always falls back on the cannon." (Had he conducted his own experiments, one wonders, wearing a tin hat on a private range out in the veldt?) "That test proves that the earth is stationary. It is our proof no. 45 that the earth is not a globe." To this he adds the Evidence of the "Shadow." "In the tropics a six foot man has no shadow at noon and for sixteen miles on each side of him men have no shadow at noon, but men farther away begin to have shadows. This test of the vertical rays of the sun indicates that the diameter of the sun is only 32 miles . . ."

This sort of thing is not like the position of men like Dr. Udintsev, the lone skeptic who doubts tectonic theory. But the point at which a serious scientist becomes a "nut" is not without interest. The original Zetetics were followers of Pyrrho, the founder of the skeptic philosophy, and their name derives from the Greek verb "to seek." They thought of themselves as searchers and inquirers, not believers. The example of "Parallax" and others like him shows how swiftly a text or collection of assertions turns into a doctrine which itself becomes a comfortable hive for every bee in every bonnet. Yet there ought to be an aspect of zetetics which confines itself to an absolute skepticism both playful and imaginative. There was an admirable movement which resolutely refused to believe in the US moon landing of 1969, maintaining that the entire thing had been a brilliant hoax designed to cow and discomfit the USSR as well as to ensure congressional votes for the further unlimited funding of NASA. The idea was that the astronauts had never left the launch pad, that the extravaganza had been created in the studio using simulation techniques and special effects and fed into millions of TV sets around the world where it was received by a credulous audience. The merit of this argument, apart from being funny, was that the average viewer could not refute it. It made the point that a TV audience will believe anything they see provided it is served up with the right trappings and couched in the approved "History in the Making" rhetoric.

It is not only from an anarchic streak that one would nearly prefer this version of events, but mainly because skepticism constantly frees thought yet can coexist with knowledge. Only at irritable moments aboard *Farnella* would I have been tempted to argue that GLORIA was an expensive hoax and the seabed at an infinite depth. Even the denial of a true idea creates a space which vibrates with possibility.

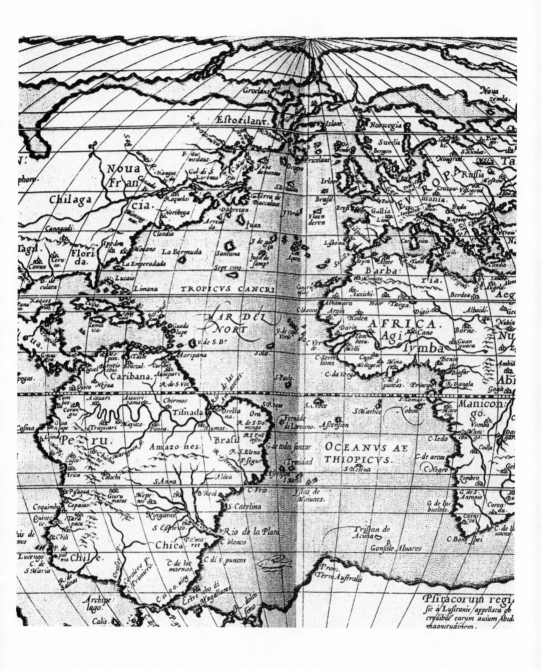

A detail from Abraham Ortelius's map of the world (1570)
showing a selection of North Atlantic islands, both imaginary
and misplaced.

This place has no name. . . . Nor does the lost swimmer even have a place, buoyed up as he is in an illimitable steep of fluid and, for all he knows, borne along by a current. In all directions is void, whether air or water, though busy with sunlight and spangles and small events. The sea itself is calm. He wishes there were a higher swell so he could more easily keep up the hope that his boat, even though not many yards away, remains hidden by conspiracies of wavelets. He knows exactly what it would be like to be in an airplane flying above where he is now. He knows the burnished pane of ocean with its frozen wrinkles crossed by the aircraft's shadow. He knows, too, how words like "millpond" only ever come into the mind when gazing disembodiedly out of a window at 20,000 feet. This leaves the swimmer with an echo from which to build a name, "Despond," for this locus in which he is adrift before he abandons it as hackneyed and unhelpful.

At last he works out that this place can have no name other than his own. Nothing if not isolate, he is himself an island. By mischance or gross carelessness he has become marooned on himself. This perception has a point in its favor. It is an island with room for only one castaway. In the almost impossible event of anybody else reaching its shores they would at least be coming as rescuers.

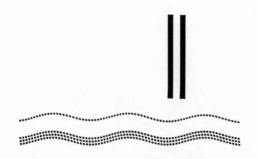

ISLANDS AND BOUNDARIES

1. ISLANDS AND BOUNDARIES

Tiwarik lies little more than half a mile off the coast, immediately opposite the fishing village of Sabay, which from the seaward side appears as a straggle of huts on stilts half lost among the coconut groves. In front of them on the stony beach boats are drawn up, each of which in time becomes identifiable so it is possible on any particular day to read the beach and know who is doing what. The island is uninhabited (uninhabitable, practically, since there is no water) and tiny, being about a quarter of a mile across. But its size on a map—and I have never seen a map largescale enough to mark it—would be deceptive, for it rises to a peak off in one corner which cannot be less than 400 feet above the sea. There are no beaches, merely that one shifting coral strand maybe a hundred yards long and facing the mainland. The rest of Tiwarik rises from gurgling boulders more or less vertically up volcanic cliffs of black rock. From one quarter there is a steep sweep of coarse tall cogon grass up to the forest which caps the peak. Seen from the strait on a breezy day the sunlight goes running up and up through this wild grassfield. It is the same effect as with young hair and similarly afflicts me with deepest melancholy, affection and pleasure.[1]

This was how I depicted an uninhabited island where I lived for a while. I wanted to account for its centrality to my life, while describing the lives of the villagers on the other side of the narrow strait, their habits of dynamite fishing and our nocturnal spear-gunning forays among the reefs and boulders. I noted some of the salient features of tropical offshore islets in Southeast Asia: their considerable discomforts, their frequent lack of drinking water, their distance from supplies of nearly everything other than brine, rocks, harsh coral sand, thorns and greenery.

Perhaps I wrote too much about beauty—at least, about the unidyllic way the island struck me as beautiful—since really I wanted to make the point that in the terms of holiday brochures it was nothing out of the ordinary and near to nowhere very special. I observed that it was in no sense a "paradise isle" as the tourist industry understands it, advancing its complete lack of water and, more especially, its lack of a permanent beach of soft white sand as reasons for asserting that at least "Tiwarik" would never suffer the indignity of being turned into a resort. Its unyielding, basaltic indifference to any melting aspect, to the dreamy topos (not a coco palm anywhere) would guarantee its own rugged persistence. Finally, to protect myself as much as the island, I gave it a fictitious name and was carefully vague about the Philippine province off which it lay.

Many months after I had finished writing the book wisps of rumor reached me away up the coast. Rumors in those parts being what cocaine is to Hollywood, I attached no importance to them. Then in due time I went back to the little fishing hamlet of "Sabay" from whose shore one can see the island a mile or so away on the other side of a strait of tearing currents.

To be the biographer of a place or person can insert a murky distance between them and oneself, especially after publication. What has been a private, even obsessive, project turns overnight into an implied claim to special knowledge or scholarship when really all one had on one's mind was love and curiosity. It becomes easy to retreat into snappish inward

[1]James Hamilton-Paterson, *Playing with Water* (1987).

pronouncement that while anybody may be a greater expert on the subject nobody else has quite the same affectionate eye . . . and up drifts the murk. But on this occasion I sat on an empty oxygen cylinder on "Sabay" beach and gazed through a clear lens of air at "Tiwarik," picking out the details which I still felt had written me as much as I them. On that very outcropping I had lived, had fought a grass fire, dried my catches, had worms, been bitten by centipedes, had watched a pair of sea eagles come and go to their nest in the cap of jungle with fish in their claws. I had glimpsed much else besides, and often felt I had voyaged on it farther than the island's small boundaries.

"No more," said a friend. "It's been sold."

A knife-like stroke. Studiedly offhand, though: "Oh? To whom?"

"Japanese. Very rich. *Ayy* . . . very big project, James. Very big plans."

"For *that*? Oh, nonsense. *Baka tsismis lang*." Just another of those rumors (narcotic, stimulant, currency). Just the favorite Filipino pastime of telling tales of projects which are going to transform the hardscrabble of living, tales which occupy a psychic territory as much as a local site. Sunken galleons . . . Japanese war chests . . . Chinese pirate hoards . . . wood-burning power stations. . . . The tales fade, reemerge, fade again. The captains and the scuba divers depart; the phrase "feasibility study" drops from conversations. Things go on being the same.

Not this time, apparently. This time they really were to change. I paddled across to "Tiwarik," half despondent and half thinking there was probably nothing to worry about. The familiar difficult strand, the familiar steep climb up the cliff path were the same. But up on what I knew privately as The Field of Crabs were signs that people other than local farmers and fishermen had been there. Ominous pegs had been hammered into the baked soil. A shallow pit had been dug and a red-and-white surveyor's ranging rod stuck up out of it like a thermometer in the mouth of a sick man. Only in the most oblique and imaginative sense had I ever thought of "Tiwarik" as my island. Now it wore the anonymous, severed look of real estate.

For an hour or two I wandered about without being able to reclaim it. I stood in favorite places and looked down at the suck and surge of water, down through water to the island's roots. Those boulders, ledges, shelves, coral palaces; those blue thoroughfares and weedy balconies: I knew them, had examined and hunted every inch of them by day and by flashlight. While their aerial map was still familiar they themselves had withdrawn, taking with them whatever it is that makes places vibrate when looked at with a certain eye. Over the horizon comes the world; the eye in distraction flickers and clouds and at once even the rocks shrink in upon themselves like touched anemones. The land reverts to a blob.

So I left. Over the next year or so stories reached me out of which I built my own fretful picture of the vanishing of "Tiwarik" and the creation of the "Fantasy Elephant Club," as the new resort was apparently to be known. The lack of beach would be no hindrance to the wealthy Japanese visitors who would be flown there direct from Manila by seaplane and helicopter. The last thing they would want to do was swim. There in the chalets where The Field of Crabs had once been they would live in the pallor of circular neon lights and beneath these coldly fizzing halos be massaged by geishas, recline on vinyl couches to watch porno videos, gamble and sip Chivas Regal and other drinks peculiar to duty-free life. . . . I hoped the fish eagles had lifted off disdainfully at the first roar of the chainsaws. Where they had nested a cement tower was apparently now going up. I envisaged an oriental folly, a stylized ninja pagoda symbolizing the martial art of Third World property development. I kept hoping that some of "Tiwarik's" old, implacable quality would assert itself: plagues of centipedes, perhaps; the belated discovery that the Japanese military had tested anthrax bombs there in 1944 and that all the soil more than a meter deep was virulent; even a Krakatoa-like eruption of the volcano opposite which had not let out so much as a squeak of steam in the last 50,000 years. . . . Then I heard that three people had been killed—laborers, possibly, but not from "Sabay." I packed my bag and went back to the village to see old friends and separate news from rumor.

This time there was a choice of two oxygen cylinders on the beach on which to sit. Several families depended for their income on catching small, brightly colored species of coral fish, bagging them in plastic with a liter or two of seawater and a squirt of oxygen, and shipping them off to Manila for export to the pet shops of Japan and the West. (The mortality rate was probably 80–85 percent, but as long as they left "Sabay" alive it was not the fishermen's problem.) Now from these empty cylinders the view across the strait was novel indeed: of an unfamiliarly shaped island girt with rows of what seemed to be white cement bungalows with red roofs. What had once been an empty stretch of water was full of small craft ferrying groups of laborers and materials. Even as I sat, a *bangka* carrying a large cylindrical water tank crawled heavily out of the difficult shallow channel off "Sabay" beach, its outriggers plowing the water instead of skimming its surface. Elsewhere along the beach were long heaps of sandbags. Clearly, two of the disadvantages I had wishfully imagined would guarantee the island's immunity to change were being remedied almost with disdain. No drinking water? Fetch it over in huge tankfuls. No beach? Take "Sabay's" across in sacks. What, then, is an island? The author of *The Island Within* surveys his kneecaps sticking up out of the water in his bathtub and considers them very much part of his personal mainland.[2] The image has a geological aptness. "Tiwarik" is as much a part of the mainland as "Sabay" is. It just happens to look like an island because the land between was low-lying enough to have been invaded by the sea. Its flora and fauna are scarcely affected by the intervening strait. The weather is that of the mainland, birds and seeds fly to and fro. From time to time people had made efforts to cultivate small patches of its total thirteen hectares, though lately this amounted to little more than occasionally cutting the cogon for thatch. The island's crown of virgin jungle is a miniature version of those vestiges still surviving in gullies and ravines high up Mount Malindig opposite.

"Tiwarik," then, is a crumb fallen from the mainland, made of the same dough and nourishing the same plants and animals. At the time I built my first hut on it the island had no economic function of its own. Yet it did form a casual part of several economies. Locals fished there and, especially when caught by sudden squalls or currents, would hole up on it until conditions improved. On ordinary days they might land and build a driftwood fire on the coral strand, toasting a fish for lunch. It was also used by travelers. The archipelago is full of migrants undertaking long and dangerous journeys in frail craft with ropy engines. Some of these travelers are landlubbers apprehensively trying as cheaply as possible to reach a city like Cavite or Batangas or Manila where they have heard jobs are to be had. But most are born boat people who give the impression of being refugees from dry land. Visayan fishermen spend weeks away from home, drifting from province to province, from one favorite fishing ground to another, catching and selling. Some of them

[2]Richard K. Nelson, *The Island Within* (1990).

claim to have no particular home but, gypsy-like, roam these central seas often with only a language, a dialect and a place of birth to give them geographical identity. Any of them at any time may haul their boats on to "Tiwarik's" little shore for a few hours to mend nets, cook a meal, calculate how much rice and fuel they will need to buy across the strait before pressing on again. Still other maritime vagrants are smugglers and pirates. Why else would a big, thirty-foot *bangka* from Romblon be carrying at least six boxes of grenades and a .50-calibre machine gun hidden beneath a nylon sail? They were the most affable of all, catching me mending a plywood flipper. I gave them cooking oil and a disposable butane lighter and in return they offered me a grenade for fishing. When I declined, saying I would rely on my speargun, they said "For self-defense, then. There are a lot of bad characters around these parts. Us, for example." We all laughed, I a little uneasily; after an hour or two they left, waving.

Since "Tiwarik" had no population and nothing to offer except a bit of dry land in the middle of a lot of water it was on nobody's itinerary and was no one's port of call. It lay at the crossroads of no particular routes, it formed no conceivable milestone in anyone's journey. Yet it was there to be used, to provide refuge or shelter, sticks for fires, corals and boulders for fishing. Or, for the reflective, it offered a place where one could hear only the sea's rinsing murmur, the cries of birds and, at night, the tiny hollow sigh of a lamp wick in its glass chimney.

Now this place no longer exists, and I need no reminding as our *bangka* noses onto the Fantasy Elephant Club's new beach. The tangle of boulders and thorns which had always hidden the foot of the cliff is gone. In its place is a concrete sea wall which at one end abuts the foundations for a small pier. Gray cement teeth stick up out of the blue water sprouting tufts of rusty reinforcing rods. On the spot where I had pitched a flapping shelter during a storm on my first visit to "Tiwarik" many years before stands an octagonal, open-sided beach house with at its center the beginnings of a circular bar surrounded by polished marble stools. At empty stone tables sit a variety of site officials—architects and engineers—waiting for a boat, as well as a blue-uniformed guard with a pistol and a walkie-talkie. No, he says, it isn't possible for me just to stroll on up and look around. This is a Japanese operation and things are run in an efficient and security-conscious manner. Why do I want to visit the island, anyway? I explain that I had once spent time here, had lived here alone, am curious to revisit it. The guard calls up, is told to wait. I sit down on a marble stool and watch relays of sweating boys stagger beneath the sacks of sand they are unloading from a *bangka*. Time passes; the guard speaks, his radio crackles back; more time passes. Finally, I and my two friends from "Sabay" are allowed to walk up to the site but are reminded that when it is finished the Fantasy Elephant Club will be exclusively for Japanese members.

The precipitous footpath is gone. To replace it a steeply curving road has been bulldozed across the face of the cliff. As we walk up we are passed by a roaring truck full of cement and trailing sooty fumes. At that moment the last vestiges of "Tiwarik" vanish in clouds of carbon. We stop on a curve where laborers are digging a trench for a power cable. They are from "Sabay" and I fished with one of them four years before. He tells me this is the site of the accident a month or so back, just before Christmas. Some boulders fell out of the freshly cut embankment and crushed three workers from up the coast. Two died on the spot, a third is in hospital in Manila and likely to die. I hope their families have been compensated. Oh yes, says my informant's workmate, they were each given 30,000 pesos (about $1,000).

Once at the top I am unable to recognize nearly all the familiar landmarks. I cannot even be certain where I saw the pegs and surveyor's ranging rod on my last visit. Much of The Field of Crabs is now landscaped and has been disguised as a miniature golf course. A tennis court is even now being surfaced and a small swimming pool receiving a first coat of obligatory blue. Individual bungalows—storm-windowed, air-conditioned, self-contained—are disposed among tasteful arrangements of rocks already planted with flowering shrubs. The centerpiece of the development ("the masterpiece," as the assistant to the architect told me down below in the beach pavilion within the latter's earshot) is the clubhouse itself, a long, low, white palace whose foyer is painstakingly decorated with appliqué designs of shells set in cement. This is apparently to contain all manner of restaurants, sushi bars, guest rooms and steam baths. At the moment it holds a good few Filipino laborers wearing the snipped-off corners of plaster bags on their heads as sunhats. A siren sounds from up in the forest, a strident wail which sends an instant image of escaped prisoners fleeing through the mind, but it is only to mark the beginning of the laborers' mid-morning break. Most unwrap their *merienda* from scraps of newspaper where they stand. A few head up toward a straggle of huts pitched against the steep, rocky slope in the tattered scrub which the lower reaches of the forest have now become. Even within the forest itself there are signs that trails have been hacked to the top. I can see no pagoda but my companions tell me that a chalet is planned for "those who want to be alone." What, then, of the pair of eagles, my familiars? I had tried not very hard to identify them, flicking through bird books for Southeast Asia, but each time I visualized them the print slipped from their wings as they soared away, unnamed. Assuredly they were gone. In one corner of their former home a level patch has been gouged and on this now stand three immense concrete legs.

"Those are for the electricity," my companion explains. "The cables will come across from the mainland."

And so, without need for bridges and causeways ("the whole point is that it is an *island*," a Japanese will tell me stiffly some hours later) the Fantasy Elephant Club is to be firmly tethered to the mainland by umbilical cords. The great weight of unsupported lengths of high-tension lines explains the size of the pylons, which will have to withstand the stress of typhoon winds. It is possible to make out three similar pylons tucked away among the shrubs on the distant promontory. Until the project was hatched, this particular corner of the province must have been well down on all priority lists for a surfaced road and electrification. Now, thanks to foreign developers, new poles have sprouted along the dirt track on the mainland and Betamax film shows are daily and nightly entertainment where before the villagers played cards and gossiped cruelly by candlelight. "Progress," as the mayoress of the municipality is fond of saying. Yet there is little risk that someone who knows the area will melt unresistingly into admiration for the foreigners' altruism. It is the province's impoverished power company which is meeting the bills for bringing electricity to the island. And over in "Sabay," where it is reasonable to suppose the developers might in self-interest have improved the villagers' inadequate water supply, I noticed the new hand-pump had been installed by local Rotarians, just as the recent hard surfacing of the sandy lot used for basketball had been paid for by the Lions Club.

Something had gone sour at "Sabay," I discovered, though I never found out precisely what. The villagers disdainfully said the Japanese were arrogant and too mean to work for at forty pesos a day ($1.20).

They claimed they could earn far more from their usual fishing activities. It was only the other villagers up the coast who were too lazy and ignorant to fish who would consider forty pesos a possible wage. . . . There had been friction right from the early stages of construction, I gathered. At any rate the relations of "Sabay" to this new world on its doorstep were decidedly *malabo:* cloudy, murky, ill-visaged. It seemed the neighboring village was to do better in terms of the newcomers' patronage since the developers were indeed sinking a borehole there for the island's drinking water. Later, I found out something else which might explain this state of affairs, since I never seriously believed my friends' protestations (no doubt cued to my own responses and facial expressions) that the island had been despoiled, their traditional lives of picturesque poverty ruined forever by video recorders and concrete pylons.

It turned out that what was taking shape on the island was only "Phase One" of a far larger project. "Phase Two" was destined for the mainland: 252 hectares which, if all the requisite planning permissions were hustled through, would comprise an eighteen-hole golf course, a clubhouse, cottages for guests, a two-storied home for the (Japanese) aged, a private airstrip/helipad, a skating rink (presumably roller rather than ice), a shooting range and a coffee shop. There was also talk of a casino, said the mayoress, although that might eventually more prudently be sited on the island. Further facilities to be offered would include windsurfing, skydiving, water skiing, scuba diving and riding. This "Phase Two" was intended to be open to non-Japanese visitors.

Since under Philippine law no land can be owned outright by a foreigner, the developers had to have a majority Filipino stakeholding. This was no problem for the corporation's president, since he was married to a Filipina. Most of the land he needed to acquire for "Phase Two" belonged to the village next to "Sabay." In "Sabay" itself the wife was on the point of acquiring—or had already acquired—thirty-three hectares of land for "family use." The "Sabayans" sense of grievance might therefore be partly explained by straightforward annoyance that they were not as well placed as their neighbors to cash in on this land bonanza. They clearly did prefer development to raw nature, but in order to be fully reconciled would have liked a bigger slice. The fact that all the land was at present classified as "agricultural," and that to change this classification required a local referendum as well as action at provincial government level, was a minor technicality. So was President Cory Aquino's foot-dragging land-reform policy under whose measures several landless tenant farmers in "Sabay" would have stood to gain a little agricultural land of their own. The whole development project was so grand and so powerfully backed that no landless "Sabayan" was likely to risk his neck by making a serious protest, still less bring any sort of legal action.

So that day in "Sabay" I take "Phase Two" to be pretty much a foregone conclusion. I also assume the villagers' fears about organized crime in the shape of the Yakuza and insidious disease in the shape of AIDS are not necessarily going to be allayed by the development corporation's president promising "to screen all his visitors most carefully for undesirable connections" (the reference was apparently to the Yakuza), nor by their lady mayor's assurance that she would "eradicate illegal activities such as prostitution" if they occurred. Several of the more traveled "Sabay" fishermen had seen at first hand the results of the turning of Puerto Galera in Mindoro into an international resort, and had witnessed once-lovely Boracay Island in Panay "go to the devil" and require thirteen full-time tourist security police. "There is a price for everything," a local official wearing a huge fraternity ring announced sagely. "Fighting, prostitution, AIDS, Yakuza. . . . They're a regrettable but inevitable part of the development of our country."

I soon lose heart on "Tiwarik" and cut short my unofficial visit. The last thing I notice is that the tree on top of the cliff from which Intoy and I had hung the highest, most dangerous and most beautiful swing in the world has been cut down and replaced with a concrete bench molded to resemble half a rough-hewn trunk. As we start back down the new road I discover the price of the island was 200,000 pesos, or about $67. "Tiwarik" was knocked down for £3,500 ($6,215). I also hear its Japanese visitors would be paying $320 a day for their accommodation, not including food and the use of facilities. At the scene of the recent fatal landslip I catch myself doing the sort of pointless sums which nevertheless insist on being done: three days' lodging equals one life.

All these details about "Phase Two" and the land on which it is to take shape are not by the way. They are central to the manner of "Tiwarik's" demise and how it has lost its existence as an island. Even as, a thousand miles away, Japanese technology struggles to preserve the single eroding islet of Okinotorishima so it may qualify for its own EEZ, the same technology has here decreated an island and is building in its place an exclusive peninsula. What with private helicopters and hydrofoils and high-tension cables, the narrow strait will have even less of an isolating effect than a six-lane highway has on an urban community. The island and the mainland together form one single project. Splitting it into two phases is merely an engineering and procedural convenience and has to do with two timescales, not two separate places. By annexing—perfectly legally—a piece of territory in another country and turning it into pseudo-Japan, Japan is in effect exporting its boundary. It is mere chance that the project, and the money, happen this time not to be Swiss or American or Chinese or German. The end result is the un-islanding of "Tiwarik."

~

"Tiwarik" is a good example of an island which has been lost to a kind of gerrymandering, to a redrawing of its cultural boundaries. Its mistake was to have been too amorphous, too anonymous and unclaimed except by a nomadic European who briefly thrust his own identifications onto it and wanted nothing but silence in return. That was selfish, no doubt. Also harmless. It would be a historical solecism to look back a century or two and think yearningly of the unpopulated, unneeded islands then littering the world's seas. Two centuries ago one's mind would have been quite different, with different wants and expectations in a completely dissimilar world. So, at least, we suppose. Yet islands have always exercised a fascination and—unlike deserts, for instance—are enough repositories of fantasy to be slightly chimerical. One seldom looks at an island without also imagining it disappearing behind a bank of fog or storm clouds which at length clear to reveal an empty ocean. It would not be a surprise and, like a dream, one might not even miss it. Until remarkably recently the North Atlantic was full of islands which have now disappeared. They constituted another type of vanishing island, one whose loss is due to the redrawing of maps, to improved cartography and navigation, but also to changed expectations.

These places hovered on the boundaries of the actual and the credible for hundreds of years. They cannot have emerged from nothing and neither have they entirely vanished yet in all their mysterious aspects. The most famous of them were Antillia (or Seven Cities), Brasil, St. Brandan, Buss and Mayda. Seven Cities Island had grown out of a legend of

Christian refugees, among them several bishops, fleeing the Iberian peninsula before the Moorish invasion in 711 and fetching up in a safe land somewhere well away in the far Atlantic. Travelers who claimed to have visited it had found plentiful church services and brought back sand which was one third pure gold dust. Not surprisingly it remained a place of considerable interest to navigators. So did Brasil Island, which had appeared on Dallorto's famous map of 1325 as a large, circular island some way off Munster in Ireland. In 1498 the Spanish ambassador in London reported home that "The people of Bristol have, for the last seven years, sent out every year two, three and four caravels [light ships] in search of the islands of Brasil and the Seven Cities."

Seven Cities was on Desceliers's map of 1546 at a position which today would be between 500 and 600 miles off New York. Since Columbus's day, though, mariners had been sailing the western Atlantic with increasing frequency and by the late sixteenth century the island had begun to shrink, moving southward and out into the mid-Atlantic wastes. A century later it vanished entirely. Brasil Island lasted much longer—embarrassingly so, since according to one source it "persisted in the mind of the British Admiralty until the second half of the Nineteenth century."[3]

The Isle of St. Brandan was even more illustrious, since it could trace its origins back to ancient Greece and Celtic mythology, both of which had stories about an Island of the Blessed. Pagan myth was converted into Christian parable by putting the sixth-century Abbot of Clonfert into the role of a lone seafarer who, committing his body to a leaky coracle and his soul to God, overcomes all manner of dangers and hardship to arrive at last in a land of paradise. It was this island which Charles Kingsley chose to be the home of the water babies, using it in a curious way. He preserved the fairy-tale aspects. "On still clear summer evenings, when the sun sinks down into the sea, among golden cloud-capes and cloud-islands, and locks and friths of azure sky, the sailors fancy that they see, away to westward, St. Brandan's fairy isle."[4] The religious connotations were still there, too, for "when Tom got there, he found that the Isle stood all on pillars, and that its roots were full of

[3]Yi-fu Tuan, *Topophilia* (1974).
[4]Charles Kingsley, *The Water Babies* (1863).

caves," which implied a place modeled securely on Gothic church architecture. Yet Kingsley then invoked paganism by identifying St. Brandan's with Atlantis, the lost land of Higher Knowledge, which to this day is still assiduously mapped by its faithful.

"The Sunken Island of Buss" was different in that its discovery was definitely recorded. It was found in 1578 during Martin Frobisher's third and last attempt to find a Northwest Passage. One of the smallest of his ships, the *Emmanuel,* was a buss (a herring smack of about sixty tons) out of Bridgwater in Somerset. During a severe storm off Greenland it became separated from the rest of the expedition. Completely lost somewhere in the region of Bear Sound, it began sailing southeastward in a vaguely homeward direction whereupon, in the words of a contemporary but secondhand account, "they discovered a great island in the latitude of 57 degrees and a half, which was never yet found before, and sailed three days along the coast, the land seeming to be fruiteful, full of woods, and a champaign country."[5] Almost at once the Island of Buss began appearing on charts. In the seventeenth century it was large and usually about 570 miles due west of Rockall. By the eighteenth century it had begun to shrink, improved navigation having failed to find it. Owing to its definitive position as well as to the documented date and circumstances of its discovery Buss was never free to adopt the other islands' strategy of scooting elusively about the ocean. Instead it shrank by sinking, for Van Keulen's 1745 chart observes "The submerged land of Buss is nowadays nothing but surf a quarter of a mile long with rough sea." Later that century and into the beginning of the nineteenth several ships searched for Buss by taking soundings but the results were inconclusive. Probably its last appearance on a map was in the 1858 edition of Keith Johnston's *Physical Atlas,* where it is a speck in the North Atlantic.

What did the *Emmanuel* find in 1578? After Buss had finally gone cartographers were left with four possibilities. Either it was a fraud from the start or else the disoriented sailors had mistaken a fog bank or ice field for land. Or was it a genuine island which had soon submerged? The fourth and most likely explanation is that the little buss from Bridgwater had glimpsed a stretch of the Greenland coast, probably in

[5]E. J. Payne, ed., *Voyages of the Elizabethan Seamen to America* (1893), p. 183.

the region of Cape Farewell. A certain amount of imagination and a vivid non-eyewitness account (Greenland is not an obviously "fruiteful" and well-wooded country) did the rest.

Lastly, the Island of Mayda survived on maps longest of all. It was usually crescent in shape and placed out in the Atlantic to the southwest of Ireland, more or less where the Porcupine Seabight is on modern charts. One of the variants of its name, Asmaida, probably points to its having originated with medieval Arab navigators. Like Brasil Island, it began moving steadily westward across the Atlantic, in 1566 fetching up in Newfoundland waters. By 1814 Mayda had drifted much farther south, on a level with the West Indies. According to one source,[6] its final appearance was on a Rand McNally map of 1906. It is anyone's guess as to what Mayda's objective correlative was: possibly Bermuda or Cape Cod or Cape Breton. It was most likely to have been dimly perceived America.

The modern world with Global Positioning System at its disposal, to say nothing of satellite pictures, may view these elusive islands of the North Atlantic with a degree of indulgent patronage. Yet for Mayda to have existed on maps from 1400 to 1906—albeit in a variety of locations—is a feat which testifies to something other than mere navigational error. It clearly served a real function. In its slow and stately disappearance over some 500 years the Island of Mayda—which no one had ever seen twice, still less landed on—in effect demonstrated that something cannot come into being without displacing something else. Mayda had to be pushed gradually aside in order to allow the eastern seaboard of America to come into cartographical existence.

We who carry in our heads an image of the Earth seen from space and have noticed on the surface of its blue ball continents and land masses looking identical to the ones in atlases cannot picture the world as it was to someone half a millennium ago. At that time the prevailing image was entirely influenced by Ptolemy's great *Geographia* (A.D. 90–168 *c.*). Ptolemy conceived the Earth as spherical; the problem he bequeathed was a lack of enough known world to cover the sphere. The whole of the globe was centered, as the name implied, on the Mediterranean. Every Greek sailor knew that once one had sailed beyond the last points of land the sea just went on and on. What particularly frightened the

[6]W. H. Babcock, *Legendary Islands of the Atlantic* (1922).

Greeks, and therefore the European mind which inherited their philosophical tradition, was the idea of *void*. The sea's void, that infinitely
dangerous blank beyond known land, was as worrying metaphysically as
it was physically. The Greeks' idealization of a static universe full of fixed
entities, faithfully reflected in their mathematics, underpinned all that
could be thought. The sea was a positive insult to this metaphysics, a
naked opposition to it. Not only was the ocean of unknown dimensions
but it was moving, unstable, in certain circumstances even breaking out
of its natural confines. How then could this fluid void be mapped? How
did one map an ocean when it was featureless? How did one represent
an *absence* of topography?

To appease this Aristotelian *horror vacui* mapmakers before 1500
resorted to a variety of devices, including that of corraling the ocean
safely within a complete ring of imaginary continents like the Zetetics'
ice barrier. Until roughly that date maps of the world were entirely
notional. Ptolemy had been rediscovered in about 1200, but *Geographia*
was only a text. The features of the globe were extrapolated from it
entirely according to the cartographer's fancy. Religious imagery was
prominent, some maps depicting three continents which corresponded
to the three sons of Noah. (The insistence that the globe assume a
doctrinal shape lasted at least into the Age of Enlightenment. The
French cartographer Robert de Vangoudy published a map of America
in 1769 showing the land as divided by Poseidon among his ten sons,
much to the amusement of Voltaire.) Other shapes assumed by the
world's land masses at mapmakers' whim were neat crosses, caskets and
the Tabernacle. Until a certain date the function of land on mappe-
mondes is to express the wish that the physical world should conform
to theological or aesthetic categories. Underlying this, though, runs an
anxious desire to frame a linked and anarchic series of voids into distinguishable oceans. When this is done the picture reverses out and instead
becomes a map of land with bits of sea in between. It is a profound relief.

Mayda and all the dozens of other islands in the North Atlantic were
the surviving, mobile fragments of conceptually necessary but imaginary
land. This is precisely Claude Lévi-Strauss's idea of the "floating sig-
nifier." He argues that in culture there is always a need for certain
concepts and expressions in order to soak up any excess of existence
which has not yet been turned into words. It is the analogue of the
algebraic concept of *nought*, which it is necessary to have before other

things can be deployed.[7] The Island of Mayda's function was to be nonexistent, to blot up an excess of vacancy, until something more solid turned up. Its poignancy is that even when it had been rendered redundant cartographers were loath to part with it. This was no doubt a matter of pride as much as sentimentality. Mayda was doomed to wander in an oceanic oubliette like a melting ice floe until being covered with the map's legend. When in some future edition the legend was moved, Mayda was found to have vanished.

Mythical or badly misplaced islands were not, of course, confined to the North Atlantic. The parts of the world remoter from Europe which were explored, mapped and named later had their own share which lingered correspondingly longer, even though by the middle of last century navigational methods were very reliable. Dougherty Island was believed for over a century to lie to the south of Australia. It was frequently reported at its given coordinates and in 1893 a New Zealander, Captain White, claimed to have sailed entirely around it, saying there was no other land for 1,100 miles. After that it vanished, but its image endured on US charts until 1932. The same went for Podestà Island, named for the Italian ship which discovered it, the *Barone Podestà*. It was sighted in the South Pacific, some 900 miles off the Chilean coast, and the ship's master wrote down its position. However, since his name was given as Captain Pinocchio it is perhaps not surprising it was never seen again, though it survived on some charts until 1936.

More inconvenient was the case of Sarah Ann Island, which ought to have formed part of the Gilbert Island group in today's Kiribati. In 1932 some American astronomers decided it would be the ideal spot from which to view a long-awaited total eclipse of the sun. The US Navy went off to have a look at the island and report back regarding the difficulties of accommodating scientists and their equipment on it. Despite searching, however, they failed to find Sarah Ann and the observations had instead to be made from Canton and Enderby islands, 500 miles to the east.

Of all such islands maybe the longest lived were—or are—a couple far down in the Pacific, southwest of Tierra del Fuego, Macy's Island and Swain's Island. These had vanished from practically every known

[7]See Brian Rothman, *Signifying Nothing. The Semiotics of Zero* (1987).

chart by 1939, but they are still there in the 1974 edition of the *Soviet Atlas of the Pacific Ocean*. It is not clear which country they belong to; nor, indeed, to which era. One of them even persists on a John Bartholomew map in *The Times Atlas of the World* (the 1986 reprint of the seventh edition of 1985).

If this seems like a catalog of earnest errors, the case of Hunter Island is one which from the start ought to have aroused suspicion but which was long treated very seriously in some quarters. In 1823 Captain Hunter of the brig *Donna Carmelita* claimed to have found an island some 300 miles northwest of Fiji. Not only did he establish its exact position, he landed and found an island which was intensively cultivated and lived on by a tribe of "highly developed" Polynesians. These natives had certain peculiarities. They all had their cheeks perforated in weird patterns and the little fingers of their left hands amputated at the second joint. For many years passing skippers kept their eyes skinned for Hunter Island, since apart from anything else the captain's account had intrigued a good few anthropologists who had never before heard of a Polynesian tribe with these curious characteristics. Alas, the island was never seen again. The idea that the entire thing might have been invented as a joke by a bored captain with a gift for subversive fantasy might have crossed somebody's mind on learning that his original—and private—name for the island was Onaneuse.[8]

~

"Tiwarik" and Onaneuse and, for that matter, all the world's chimerical islands have things in common which do not depend in the slightest on notions of "objective reality." To the gaze beneath which they once fell they had an absolute existence. St. Brandan's exercised its significance on the religious imagination of the time, while Seven Cities with its gold sand and Buss with its "champaign country" stood as lands of promise whose only fault was to slip further out of reach as the centuries passed. Onaneuse, since all jokes are serious until their inventors begin to laugh, maybe stood for something which had been at the back of Captain Hunter's mind long before he ever went to sea. It might be oddly reassuring to invent a private erotic idyll, give it precise bearings, and

[8]See Karl Baarslag, *Islands of Adventure* (1941), and Henry Stommel, *Lost Islands* (1984).

then watch other people search for it in vain. While the captain laughed, the scholars accorded it a status he would never have dared allow himself to claim. They did his work for him even as they failed to find it, and since he had known all along it wasn't there he had the last laugh as well.

No doubt islands draw some of their peculiar significance from the dozens of cosmogonies which begin with a watery chaos out of which land emerges. Any emergent land must initially take the form of an island, so the island stands as the archetype of land. As to what this proto-land might contain would depend on when it was first spied, and by whom. Paradise, treasure and naturalists' nightmares were variously seen as appropriate, but the nearer our own century was approached the more an explorer, an adventurer or a philosopher might expect the proto-land (glimpsed tantalizingly in the parting of a fog bank or glittering in the objective of his telescope) to contain a domestic order reassuringly old-fashioned as well as exotically unlike any known society. Things are different nowadays. Nobody any longer expects to find a place where the people are nobler, sexier or just better behaved. Wistfulness has been replaced by a certain hard-nosed quality. If you can't find them, you found them. Were there such a thing as an endangered species of land, the island would be it. Far from being proto-land it is coming to feel like a last land. The whole concept of the island, which until recently was implicit with all manner of promise, is now redolent of loss.

"Tiwarik" will go on existing only as long as its author. Unlike the island to which I attached the name, it is not contingent on Japanese developers. Somewhere its grasses still blow in the wind. Six or so years ago, when it was its old self, I ended my description by calling it an act of the imagination. This will always be true of places which at last become properly real to us.

It is not its grasses my feet have trodden nor its little coastline I have so lovingly followed, and neither does it retain any trace of me. There is another island locally known as "Tiwarik" but it is only an exact facsimile, a fly-spit on the map of the objective planet which we agree to inhabit.[9]

[9]Hamilton-Paterson, *op. cit.*, p. 263.

2. OBJECTS OF DESIRE

Four things about small islands:

1. They look like objects, and hence like property.
2. The concept of the "private island" satisfies most people's major fantasies.
3. The effect of islands is almost wholly regressive.
4. An island's boundaries can never be fixed.

1. They look like objects, and hence like property.

Everyone looks at an island, whether consciously or not, much as a tyrant eyes a territory. It takes a long time to have any relationship with a land or a country, but the mere sight of an island from an aircraft's window or a ferry's deck mobilizes the beginnings of possessiveness. The place is small enough to treat with, to become familiar, to exhale an air of exclusivity, even if it is quite nondescript. A slight grammatical shift can mark either social desirability or small size—usually going together. Thus, one has a house *in* Malta, but a bungalow *on* Gozo. He lives in Jersey, she on Sark. (But they have a house on Long Island as well as one in Jamaica.)

This unit of land which fits within the retina of the approaching eye is a token of desire. The history of the Isle of Buss shows this desire working so strongly that successive mariners appropriated a portion of a long coastline and changed it into the island they would have preferred to discover. To have happened upon an unclaimed continent while lost in a small fishing smack would have been inconvenient, but to have found an unknown island was both manageable and enviable. How, then, could its discoverers have extrapolated a self-contained shape from a length of coastline? How were they able to draw the fictitious "back" of this "island" which remained forever as hidden and theoretical as the dark side of the Moon? Medieval cartographers often solved this problem by giving the Atlantic islands stylized shapes: circles, clover leaves, rectangles and crescents. The Isle of Mayda retained its crescent or indented circle shape on map after map, and eyewitness accounts of it seemed to conform to this outline with remarkable faithfulness. Quite possibly this reflected its rumored Islamic origin.

There for the taking . . . Ever mobile, for several hundred years the lost islands of the Atlantic might bob up anywhere from behind freezing mist, in a hurricane, or during a search for somewhere else entirely. The point was they could be possessed at the drop of an anchor, named for a vessel, claimed for a monarch. Even today, visitors and holidaymakers may "discover" an island which becomes "theirs" in respect to their friends, envious neighbors, peers.

Icebergs, floes and ice islands also form a particular class of islands in that they are both mobile and temporary. They look like and are objects, and are sometimes colonized by Eskimo hunters and teams of scientists for varying lengths of time. Several islands made of shelf ice and far larger than the Isles of Wight or Man have provided stable bases for research stations for ten or more years at a stretch. The question of possession is another matter, though. If they are inside territorial waters there is no problem; but many icebergs carry with them a vast tonnage of boulders and other morainal material and one might wonder to what extent they go on being part of the nation in whose territory they were calved. Canadian soil and Canadian water presumably made up the iceberg which sank the *Titanic;* but while the Canadians would have retained full rights over it while it was in their waters, would they automatically have ceded all responsibility once it had left? Presumably so, otherwise the White Star line could possibly have brought an action against the Dominion for negligence in allowing pieces of its sovereign territory to go drifting away out of its control.

The boulders carried by such icebergs and released as they melt often end up thousands of miles away from their place of origin and in the early deep-dredging oceanographic expeditions of the mid-nineteenth century caused geological confusion. Were such boulders now discovered to contain valuable rare minerals in exploitable quantities, perhaps Canada and Denmark might pursue a legal claim granting them exclusive mining rights over their pieces of rogue territory shed from Newfoundland and Greenland, a claim which would also exonerate them from blame when anyone accidentally rammed one of their melting assets.

2. The concept of the "private island" satisfies most
   people's major fantasies.

This assertion may be true only for our culture; but as Western culture in general seems regrettably set eventually to subsume most others it is probably at worst a truism. The "private island" fantasy is simply one expression of the urge to define, annex and defend territory. It is clear that in this context "island" can as easily mean any patch of land anywhere, even a mere house. This is especially noticeable in England, where his home is only half-jokingly referred to as "the Englishman's castle." The apotheosis of this is a place like Loch Leven Castle in Kinross, where Mary, Queen of Scots was imprisoned and from which she escaped in a rowing boat. This consists of a castle on an island in the middle of a lake which is itself on an island in the NE Atlantic. The idea embodied in this arrangement can be expressed graphically by a series of concentric rings: circular boundaries nesting within one another, lines of exclusivity and defense which intensify in power the more they approach the center. The average medieval castle, in default of a handy lake with an island, had to make do with a moat, thereby becoming an artificial island.

There is certainly a tendency, perhaps more pronounced in some cultures than in others, to make "islands" on dry land. In Tuscany, for example, the natives increasingly resent the habit of foreigners (meaning both non-Italians and non-Tuscans) of buying up pieces of their countryside, fencing them off and forbidding all access to them by locals. As with the villagers of "Sabay," the resentment is mainly twofold. The Tuscans do not like their immemorial rights to hunt, gather, stroll or otherwise come and go suddenly abrogated, nor what they have always considered part of their horizon to be out of bounds to them. At the same time they are put out that it is not they themselves who had the financial liquidity to take advantage of the boom in local land values. Had they done so it is open to anyone to wonder whether they might not with alacrity have assumed the grandiose mantle of landowner, and as swiftly put up fences and given large, mean dogs the run of their property.

The "private island" remains the correlative of a particular dream. Islands are at once objects of desire and a locus for desires. The dream embodies fantasies of autonomy, independence, security, sex, grandeur, individuality and survival, in recognition that modern metropolitan and suburban life connotes powerlessness, dependence, defenselessness, frustration, lack of status, anonymity and a general feeling of expendability. In waiting rooms people eye color advertisements in *Country Life*, aerial views of yet another Scottish island about to come under the auctioneer's hammer, while an easily decoded dream crosses their mental retinas and glazes their eyes. *Estimated price: $1,316,300.* The same dream leaks into all sorts of stories and films set on private islands where the unities of time and place can be rigidly controlled. These may be tales of manhunts with the narrator-guest as the next quarry; reigns of terror; ghoulish experiments; masterminds plotting the world's overthrow from their flamboyant yet top-secret lairs; elaborate erotic baroqueries. Science fiction carries the dream on, being full of expansive futures in which the rich and powerful own private planets, while even the moderately wealthy may aspire to a humble asteroid as the site of a kingdom, retreat, hideout or love nest.

Nor is the dream confined to adults. In their coastlines, as in their potentiality, all lost islands go on reappearing in the maps which every powerless schoolchild draws.

3. The effect of islands is almost wholly regressive.

Islands infantilize people even as people idealize islands. Those with appetites and no souls think they would be safe from the eyes of the world. Those with Soul and little appetite believe they can fall under an island's benign and teaching gaze.

The island repeats a fantasy of human beginnings. The fetus—castle of the ego and keep of the soul—is effectively an island for the first nine months of life, entirely surrounded by an amniotic moat and connected to the mainland only by an umbilicus. Soon afterward the playpen becomes an island, probably the most fabulous of all. Not only does the infant command its every square foot, he commands the world which his own supreme frontiers deign into being by marking off. His shores, his limen; and so by extension his ocean, his continents, his world.[10] More-over, the fantasy of a private island always takes on that infantile charac-teristic of absolute flexibility in being able simultaneously to stand for almost any desire and to serve as the ideal locus for practically any fantasy. For islands are also sexual places because they have the air of being extralegal, extraterritorial, out of sight and censure. Every so often a film appears depicting torrid intimacies among the conveniently ma-rooned. For this cinematic purpose the island must be tropical and the state of undress constant. It would not be at all the same for two nubile castaways to find themselves stranded in the Bering Sea.

The island is thus the perfect territorial expression of the ego. As such, it is all too easily a metaphor for the individual. Sometimes the metaphor is used at one remove, so the island takes the place of a wise *alter ego*. The message here is that man learns by true experience of himself. The lessons may be practical and moral (as in *The Swiss Family Robinson* or the story of Alexander Selkirk) or spiritual (as in Richard Nelson's *The Island Within*).

The infinitely flexible nature of islands, of their being at once safe and adventurous, constraining and boundless, erotic and polemical, has

[10]Part of the island's haunting quality may be because its exclusivity reminds us of the family as we once saw it through infant eyes: self-contained and self-sufficient. A family's underlying sadness resides in its conspiracy of immortality. When decades later we come to look at it with an almost-stranger's eyes, a family relic such as an old tablecloth now stands poignantly revealed in its faded colors and moth holes as having always been both altar cloth and shroud.

made them ideal destinations in a long literary tradition of imaginary voyages. More than a thousand years before Homer there was a twelfth-dynasty Egyptian story about a castaway on a marvelous island, and Plato's account of Atlantis functions as a kind of blueprint on which he might later have constructed a more complete utopia. When Sir Thomas More produced his own original *Utopia* in 1516 he put fresh life into an ancient genre. The dignity of his Latin must have induced many a lesser writer to indulge his own intellectual fantasies under the disguise of gravity, for the literature of the next three centuries abounds with all kinds of utopias and ideal commonwealths, most of them sited on imaginary islands. (At this point, and quite gratuitously, I wish to note an allegation that Sir Thomas More "used to thrash his grown-up daughters with a rod made from *peacock-feathers.*"[11] Without bothering to try and put a finger on it more precisely one feels this sort of behavior is not inconsistent with thought about islands and ideal societies.)

It is curious there was no discussion in English of the imaginary voyage as a genre before the nineteenth century. Indeed, there was not even any recognition that it was a literary type worth discussing. In France, on the other hand, there were all sorts of studies and by 1787, when Garnier's remarkable *Voyages Imaginaires, songes, visions . . .* was published, he was able to subdivide his classification of Allegory into a whole variety of islands, among them an *île d'amour,* an *île de la félicité,* an *île taciturne,* an *île enjouée,* an *île imaginaire,* and an *île de portraiture.* After Crusoe's great success in France, several imitative *Robinsonades* showed what man might be capable of when thrown entirely on his own resources, whereas adventures on an *île inconnue* tended to depict what happened to a domestic society cut off from the rest of the world. In this respect they constituted something of a counterpart interest to that in feral children (such as Victor, the wild boy of Aveyron), around which at that time all sorts of arguments revolved concerning what exactly constituted "the natural" and "the civilized."

The genre still exists. The French writer Georges Perec in a novel published in 1975[12] uses an imaginary island off Cape Horn as the setting for a fascistic society obsessed with sports. And what else is one

[11] Magnus Hirschfeld, ed. Norman Haire, *Sexual Anomalies and Perversions* (1959), p. 396.

[12] Georges Perec, *W ou le souvenir d'enfance* (1975).

kind of science fiction but a convenient locating of utopias and dystopias off-Earth on imaginary planets which are, from our perspective and by any other name, islands in space?

The effect of islands is almost wholly regressive . . . This is most true of prison islands, since there is nowhere more regressive than a prison. "Regressive" should not, of course, be read as a synonym for "comforting," although many brutalized and institutionalized people find incarceration reassuring. The objections are obvious. Did Napoleon "regress" on St. Helena? What was so comforting about *le bagne* which induced "Papillon" to escape Devil's Island? Was the camp on Blood Island in some way deeply cheering? But "regressive" applies as much to the behavior of a prison's governor and guards as to that of its inmates. It refers to the effect on everybody concerned which all institutions exercise and penal institutions in particular. Punishment is by its nature regressive and prisons usually involve extremes of pettiness, brutality and sexual license: the normal ingredients of infantile behavior. Seen in this light, Devil's Island was something like a nineteenth-century English public school with mosquitoes and a guillotine. A succession of the prison's governors encouraged their charges to settle into some variety of homosexual marriage, jointly tilling a small garden and sharing its produce. This was considered an effective antidote to the yearning for escape, a progressive piece of penology known locally as "the cucumber solution" (*la résolution du concombre.*)[13]

At the very center of imprisonment's concentric rings is solitary confinement: isolation (from *insula*), which if inflicted for long enough on the wrong individual may cause a regression from which it can be hard to emerge. It all depends on one's position. From the authorities' point of view, isolation is the place where the community expels the individual. From certain individuals' points of view, solitude is what they long for to escape the community. Their yearning may even be for the numinous, invisible spot in the center, identity's apotheosis and vanishing point.

[13]See Chapter VII, note 13.

4. An island's boundaries can never be fixed.

Chapter I hints at some of the problems international law has in trying to fix Exclusive Economic Zones. In terms of absolute position, boundaries, like the Earth's crust itself, can remain fluid. Thus when I hypothesized the case of an island's EEZ happening to cross the edge of a tectonic plate and therefore either widening or shrinking, the response at IOS was a scathing "Good God, they can't measure to the nearest nautical mile, let alone to the nearest centimeter." Over time, though, these centimeters add up. In 1492 when Columbus crossed the Atlantic, America was twenty meters nearer to Europe than it is today.

One may multiply almost indefinitely the special cases of volcanically inconstant islands, islands whose coastlines are hard to define, artificial islands and self-proclaimed independent principalities sited on World War II antiaircraft towers six miles off the Essex coast (a perfect example of an island as egoic headquarters).[14] There are islands like Rockall which seem to have no purpose but to act as a focus for legal wranglings, their sole value being as pegs in the ocean around which a lucky winner might draw an EEZ. Japan has recently been driven to considerable feats of technology in order to rescue its own eroding islet of Okinotorishima at the extreme southern tip of the Japanese archipelago. At high tide Okinotorishima consists of two lumps of coral respectively three meters and five meters wide. Without these little rocks Japan's territory would end at Iwo Jima and she would lose 154,440 square miles of EEZ and with it all exclusive fishing and mineral rights. With each typhoon the rocks erode a bit more, so the Japanese have spent upward of $300 million to enclose the island in a concrete and steel protective wall. It is important that no part of the wall should touch the corals because this would turn Okinotorishima into an artificial island and disqualify it from having an EEZ. As a professor of international law, Soji Yamamoto, says: "Territory is not something that exists naturally. It must be obtained by a country's efforts."[15]

Meanwhile, if sea levels rise as predicted over the next half-century,

[14]"Sealand," otherwise known as Roughs Tower, turned by Major and Mrs. Roy Bates into a "state" with its own constitution, flag, stamps, coinage and passport. See *The Independent*, 24 February 1990.

[15]See *Nature* 333 (9 June 1988), p. 487, and *Fortune* 122 (24 September 1990), p. 12.

what of island boundaries then? Will the original outlines of a largely submerged Maldives be maintained on maps as much out of respect as for legal or navigational reasons, so sightseers can peer down from the rail of a boat at a lost land as they might today at a sunken battleship?

Other kinds of island boundary are blurred, too, especially when there seems to be no logical or geological reason for assigning a particular nationality. Corsica is historically, linguistically and geographically more Italian than French. The Channel Islands might quite as easily be French as British, and with their degree of autonomy form part of that odd, anomalous category of offshore tax havens which are neither fully integrated nor fully independent. The Isle of Man also falls into this category. Such islands generally have a certain diehard quality and are as much leftovers of an older social order as they are of a former geological configuration.

It is archipelagoes and chains of islands which are so often the geographical versions of displaced persons, holding at best a temporary passport. The Sulu Archipelago is a perfect example. One has only to sit on the wharf in Jolo to be prey to a literary sense of unreality. The waterfront of huts built on stilts over the sea, the lumps of islands at every distance, the decaying ferries and wooden launches full of fish and copra and red logs: allowing for a lack—but not complete absence—of sail, it is Conrad's horizon still, and filled as ever with the dreamy tropic energy which slops across all boundaries. Politically, Sulu is part of the Philippines right down to Tawitawi, at its closest reaching to within a few miles of Sabah in northern Borneo. However, it was recently declared an autonomous region with special barter-trade rights between it and Malaysia. This was in response to the long and bloody war waged by the Moro National Liberation Front against what they thought of as an attempt by Christian Manila to oust or dilute Islamic culture and greedily expropriate whatever it is that governments habitually do greedily expropriate.

On Jolo jetty Conrad's azure map lies ahead, while immediately behind is a troubled, dark green backdrop of conspiracy and heavy weapons. Conrad would have recognized that, too, since men here have always gone over-armed. A drunken fight can lead within minutes to mortar rounds. The Republic of the Philippines, with its implied promise of centralized law and institutionalized order, covers Sulu with a cartographer's fiction. The islands of the archipelago are defined and individuated by language, usage, tribal politics, gangs, bandits, even pride. They are crisscrossed by the interests of disparate ethnic groups, trading links, smuggling, piracy, local tyrannies, fishing, seditious movements and intersecting anarchies. In such places official boundaries vanish entirely unless drawn fleetingly by the wakes of Navy patrol craft or coastguard cutters. I once went on a week's fishing trip in an open boat from Palawan southward. We fished for lobster off Bugsuk and Balabac islands, sleeping at irregular intervals wedged into the bows or on occasional dry land. I lost all sense of time and position. On the way back I discovered we must have spent one night on Borneo. The same thorns, mangroves and littoral clutter, it had seemed nowhere different.

The example of "Tiwarik" and its grandiose conversion from nondescript islet into businessman's fantasy, vaulting a strait to appropriate a chunk of mainland, is a reminder that if physical and legal boundaries can often only be fixed with great difficulty, then areas of wish can never be clearly demarcated. The greed of dreams is to expand into any space denied them. Fantasies, daydreams and dreams flap about as they may, but they all roost in the unconscious and share its logic, which is that there are no contradictions. Just as sexual fantasies can simultaneously involve a single person and many, at once watching and being watched, doing and being done to, so may an island be experienced as both small and infinitely large, part of the land and part of the sea, sheltering and exposing, terra firma and freakishly unstable. As they melt, icebergs may become very unbalanced, often rolling completely over without warning. The private nickname "Tiwarik" was prescient, for the word means "upside down" and something about the place had always suggested rich possibilities of inversion. Never had I thought to see it so thoroughly stood on its head.

~

All this points to a boundary which as yet appears on no map but which is ever more real and ever more cavalier. If one considers ex-"Tiwarik" as a piece of terrain which has recently been annexed for the convenience of a foreign "lifestyle" then probably there can be no such thing as a holiday island. Once an island becomes a resort it ceases in some essential way to be an island and turns into an extension of a mainland, even if that is half the world away. Nor need this mainland even exist as a sovereign country. It is enough to comprise that fictional international place to whose citizenship so many lay claim or aspire: that stateless state of BMWs, Chivas Regal and Dunhill lighters; of treasure and pleasure and leisure, guaranteed moth- and rustproof. Since much of this fabled place lies in chilly latitudes it needs to push out long, tentacle-like peninsulas into warmer climes: Vacationland, embracing all sorts of otherwise underused bits of lesser countries in the sun. It simply kits these out with its own standard furniture (scuba gear, water skis, hang gliders, beach barbecues, rock music, drunks, whores). And suddenly there are no more islands, only scattered slabs of a single moneyed empire joined each to the other by something solider than water. Or so it seemed, waiting for "clearance" in the new beach pavilion on what was once "Tiwarik." This building is now a border post. Under Philippine law no private individual may own any part of the foreshore below a point thirty feet above the mean high-tide line. It is national property. Hence anyone may land on the Fantasy Elephant Club's beach. But where once by civilized common usage they were free to wander the rest of the island, they now have to be vetted by an immigration official in a blue uniform and wearing a gun. The beach has changed from a haven into a frontier.

There is one last kind of island, one whose elusive presence flickers at the edge of vision, quick as fish. This is the imaginary island faithfully mapped in every psyche, mostly unsuspected, infrequently discovered, even more rarely inhabited. An outcropping of the self, it lies across a treacherous strait which discourages acquisitiveness, and even on clear blue days may have vanished as if it were roaming the oceans in search of the one worthy inhabitant. Then on a rare day the rare person wakes and it has swum out of the corner of his eye and stands before him. On such a morning it takes no effort to cross over, paddle flashing in the sun, until the skiff's bows nudge grindingly into the shore.

And then what pleasure to set up a hut, a fish drier; to pare things back to water and light, to knives and spearpoints, to order and silence! *All men have an island,* Donne should have said, for a suspended wheel rim being beaten in a cement-block chapel on the distant mainland ought to tell us no more than the fish curling and flapping between our hands, bleeding rusty threads into the sea. That steely tolling from across the water brings no news, nothing we do not already know as later we climb the headland to watch soft dusk well up over the world's rim and efface the ocean below. It is not interesting to tot up the sunsets seen and perhaps to come. Those deaths, our deaths, are not plangent affairs but matters of geology. We are all at best marginalia in another era's fossil record. Go down to the hut instead through a drift of fireflies. Light the lamp, cook rice. There is nobody else on this island; there never was and never could be. Outside, the waves wring green flashes from plankton. The great mineral machine turns its fluid gears. The firefly in the thatch tugs us into its gravitational field.

Did Britain cease to be an island at some debatable moment between the end of World War II and the construction of the Channel Tunnel? Until 1939 Britain had lavishly exported its boundaries in terms of Empire, and while much was made of the idea of this "tiny island nation" ruling a good part of the globe, it did still retain certain characteristics of an island. Among these were the world's largest maritime presence, a thriving fishing industry around its entire coastline and an attitude toward the rest of the world which can only be described as insular. (The putative headline THICK FOG—CONTINENT ISOLATED puts it quite succinctly.)

The war provoked insular rhetoric, ranging from endless Shakespearean tags about "this precious stone set in the silver sea" to reemphasizing the tininess of "this island nation," although not now in comparison with the world it ruled but with the Hitlerian might threatening to crush it as it "stood alone in the dark days." The white cliffs of Dover stopped being rather striking beds of calcareous plankton and took on the quality of Britain's boundary, ramparts, bulwark, palisades. A few miles of cliffs came to stand for an entire spiritual seaboard which might not be violated. Had that portion of Britain's coastline most adjacent to the rest of the world been as low and unremarkable as it is in parts of Essex, Suffolk or Lincolnshire, the rhetoricians and songwriters would have needed to come up with a different trope altogether.

After the war, Britain with its depleted spirit and collapsed economy became vulnerable to the very things which erode islands: links with foreigners and—as the economy improved—increasing travel abroad by the islanders themselves. Somewhere between 1945 and 1990 Britain lost consciousness of itself as an island. The sea which surrounds it (and which only poetic fancy could ever have described as "silver") now plays virtually no role in its thinking or its economy. Its naval and merchant fleets are shadows; its fishing industry is ravaged; even its own people find the beaches of Spain and the Greek islands more congenial than those at home. So when the two pilot tunnels, one from England and one from France, met beneath the Channel in 1990 and Britain became technically joined to the rest of Europe by dry land, the event seemed to provoke remarkably little upheaval in the nation's psyche. King Arthur, Drake and Churchill slumbered on, unmoved. The country was already part of the EC, with the free movement of member citizens guaranteed. Insular attitudes still abounded, but something had changed. A generation had grown up which knew and cared nothing of precious stones set in silver seas and would in any case think the description fitted Barbados or Mykonos far better.

It is probable that only a war which threatened Britain directly could ever resurrect the insular rhetoric. In his speech in *Richard II* John of Gaunt did mention infection, however ("This fortress built by Nature for herself/Against infection and the hand of war . . ."). It so happens that Britain does still jealously guard the ghost of its island self by means of a last and potent symbol of foreign contamination: the fear of rabies. Ironically, it is mad dogs which give the Englishman his final solitary status.

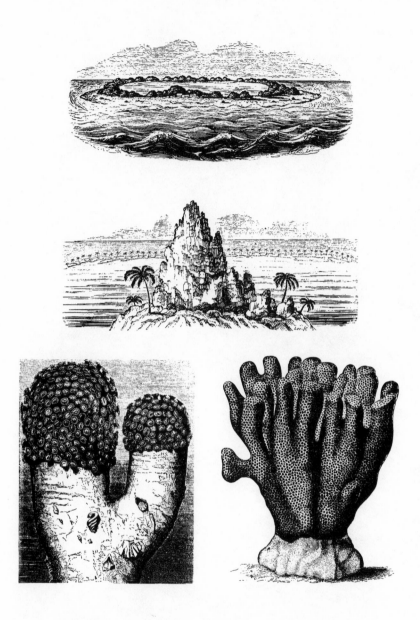

Figures chosen by Darwin to illustrate his theory of coral reef formation:
(i) A small island (Bolabola) surrounded by a barrier reef
(ii) A "lagoon island" or atoll whose central peak has sunk, in this case (Whitsunday Island) leaving a rare example of a complete ring of land.
(iii) *Goniopora columna*
(iv) *Porites mordax*

There is about the swimmer a sentimentality, or self-pity, which disgusts him even as he finds himself thinking that surely he ought to have earned a reprieve. All that close attentiveness to the sea over the years, to this ever-yielding, stony-hearted medium which has him in its embrace—it cannot have been wasted. There must be something he has learned from it, some subliminal message from his ancestral home, instructions for survival. The idea is fatuous but persistent.

He is beginning to tire. Not of staying afloat, since this is effortless, but of trudging the water to stand higher, of spinning to keep every horizon constantly in view. More and more he allows his face to hang in the water. Through the glass panel of his mask his vision lengthens past the rope's end twenty feet below his ankle. He no longer sees the prismatic chips of phytoplankton, the blazing motes and jellies as they drift past his face. Now he believes he glimpses shapes far beneath, not predators but bulks of deeper purple as though . . . Why not? This archipelago is full of hidden reefs, its contorted seabed thrusts up unexpected pinnacles to within feet of the surface in the middle of nowhere. There could easily be a coralline peak, a ledge, even a plateau over in that direction away from the sun where the water does seem to deepen its color as if a little farther down there lay a solidity. . . .

Away from the sun? The swimmer jerks his head up, gasping and squinting painfully at the blazing disc overhead. Is it not past its zenith now? Has it not begun to sink? May this illusion of a darker bulk be nothing but his own shadow, cast as he has so often seen it in late afternoon? No; ridiculous. It is not late afternoon, merely maybe a few minutes past midday. He looks downward again and in a while his eyes adjust from the dazzle and once more he thinks he can pick out an area of deeper tone. So convinced is he that he begins swimming toward it, slowly, so as not to give the impression of having finally picked a direction or of expecting very much. Now and again he glances around to tell his invisible boat where he is going. It seems to him that all will be well if it turns out to be a lonely reef he is heading for. However small, it will convert the ocean at that point into a shallow sea. He would then, as if by magic, be in his depth in twenty feet of water. Or at least, hovering as if in air above the unknown but familiar city, he would be close enough to feel its dwellers might intercede for him, present his case for survival at some court of marine jurisprudence. . . . As an idea it is better than nothing; even a glimpse of a reef would sustain him. Besides, reefs were mysterious and deceptive places whose greater being remained hidden. If one kept one's eyes on a reef under water and followed its avenues, it had a habit of turning into a shore. The swimmer had experienced it a thousand times. Might not this one, despite an apparently yawning horizon, somehow work the same friendly trick?

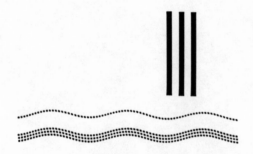

REEFS AND SEEING

1. REEFS AND SEEING

Coral was used variously as jewelry, medicine (ground up like mummy-powder) and inlays in cabinet-making. The precious red variety was traded for thousands of years without anyone being able to say precisely what it was. It was found in the sea, certainly, but fishermen and dealers were studiedly vague in order to protect their source. It presented a problem in the eighteenth century, however, since with the development of natural history it finally became necessary to say whether it was animal, vegetable or mineral.

Perversely, it exhibited characteristics of all three. At a time of industrious taxonomy this was unsatisfactory and exasperating. Coral reefs were notorious from the lyrical and often ambivalent accounts of sailors, who could one day view them as luxuriant herbaceous borders of surpassing delicacy and beauty, and the next as the most treacherous threat the tropical oceans could offer, ready at the first sign of a dozing lookout to tear the bottom out of a ship. The imagery used to describe reefs reflected this ambivalence; it also reflected confusion about the substance being dealt with. Horticultural images were common, and R. M. Ballantyne's view of a reef in his novel *The Coral Island* (1858) as a sublime rock garden was by then almost a cliché, one of the two commonest ways of seeing coral. The other, somewhat later, was to view it as architecture. A reef was a fantastical cityscape, habitation for King Neptune and his courtiers (that is, when they were not housed in an Atlantean version of this kingdom, whose style tended toward Hollywood Athenian). Perhaps certain illustrations of science fiction cities, in their lack of any unifying architectural style and their proliferation of bizarre organic shapes, derive ultimately from the iconography of reefs.

Until about 1750 zoophytes—that is, plant-like organisms such as anemones—were classified as vegetable. The very terminology (animal-plant) reflected uncertainty. At that time "the animal" was distinguished from the other class of living creatures as having the capacity to move at will from place to place. It was a long while before the agents responsible for coral were identified as zoophytes; but when they were, many saw them as vegetable, since like marine plants they were attached to the substrate by disc or tubular fibers. What was more, the branching arms of coral looked convincingly arborescent.

This classification was challenged by mineralogists. They argued that many zoophytes were hard and stony, a description which fitted no known plant. Clearly they were rocks, made of calcareous and clayey sediments "moulded into the figures of trees and mosses by the motion of the waves, by crystallisation, by the incrustation of real fuci [seaweeds] or by some imagined vegetative power in brute matter."[1] It was not for nothing that many reef-building corals became known generically as "madrepores," or mother of stone.

The animal theory had its supporters, too. According to one of its proponents, Dr. Vitaliano Donati, the "stone" of coral was not stone at all, but *bone.* "I am now of the opinion, that coral is nothing else than a real animal, which has a very great number of heads. I consider the polypes of coral as the heads of the animal. This animal has a bone ramified in the shape of a shrub. This bone is covered with a kind of

[1]George Johnston, *A History of the British Zoophytes,* 2nd ed. (1847).

flesh, which is the flesh of the animal."[2] It is an ingenious and eccentric idea. An animal in the shape of a shrub is—if nothing else—the exact opposite of topiary, and suitable for a zoological garden.

When reviewing these conflicting theories one might imagine the issue could have been settled once and for all by any naturalist spending a few hours with a good microscope. Yet a certain Henry Baker, in his book *Employment for the Microscope* (1753), could only see evidence for corals as mineral. According to him they were accretions of salts leached out of nearby rocks by seawater and bonding themselves chemically to "the calcareous and saline particles" of the water. At the same moment a London merchant named John Ellis began to use his microscope and come to different conclusions. Ellis was typical of a very English breed, the amateur naturalist, which was to flourish so valuably in the nineteenth century. Men with underused intelligences and hours to fill would turn their attention to the fledgling sciences where there was still an immense amount of basic taxonomic work to be done, work which any dedicated amateur in quest of knowledge could perfectly well do. Country parsons were particularly given to this sort of hobby and alone accomplished fieldwork and classification in a dozen disciplines including meteorology, geology, entomology, botany and ornithology. In addition to attending lectures at the Royal Society, Ellis, too, had a hobby. This was making miniature landscapes on sheets of paper with a collage of delicate fragments of seaweed and corallines. The more he used coral the more he wondered what it really was, finally putting it under his microscope. In 1752 he addressed the Royal Society, announcing "that these apparent plants were ramified animals, in their proper skins or cases, not locomotive, but fixed to shells of oysters, mussels, &c., and to Fucus's."[3]

Carl Linnaeus, the celebrated Swedish naturalist and father of modern taxonomy, was impressed by Ellis's work but nevertheless decided that zoophytes occupied a classificatory niche of their own "between vegetables and animals: vegetables with respect to their stems, and animals with respect to their florescence." Ellis would have none of this and wrote to the great man that he could not reconcile himself to "vegetating ani-

[2]V. Donati, *Phil. Trans.* (1757) abridg., xi, p. 83.
[3]John Ellis, *Essay on the Corallines of Great Britain* (1755), Introduction.

mals." He had a powerful sense of how things might partake of the characteristics of other things without fundamentally changing their nature. "Artful people," he wrote to Linnaeus, "may puzzle the vulgar, and tell us that the more hairy a man is, and the longer his nails grow, he is more of a vegetable than a man who shaves his hair or cuts his nails; that frogs bud like trees, when they are tadpoles; and caterpillars blossom into butterflies . . ."[4]

In the mid-eighteenth century the Royal Society in London was to some extent the international arbiter of such things, and Ellis's view won more and more endorsement as correct. Many of the naturalists of the period had tried burning corals and found they obtained much the same smells and residues as from burning animal products such as skin, hair and bones. This seemed conclusive. After Ellis there was broad agreement that zoophytes, including corals and sponges, were indeed animal though there were some who, with Linnaeus, hedged their bets by preferring to think of them as "intermediate beings partaking of a twofold nature."[5]

This version of the dispute over corals turns out to be a very Anglophile account, implying as it does that it was all a matter of a handful of Britons working more or less in isolation. Such is not the case. Recent research makes clear it was a subject much discussed elsewhere in Europe a good half-century earlier, especially in France and Italy.[6] It reveals that it was actually a Frenchman, J.-A. Peyssonnel, who first identified the living part of corals as animal. In 1706 the Italian Luigi-Fernando Marsigli (1658-1730) arrived in the French town of Montpellier, where he found "a thriving scientific community, serving the ancient university with its famous school of medicine and botanic gardens, a school of hydrography, and a Jesuit college."[7] Having collected corals, he performed all kinds of experiments on them. He would take branches home and put them in seawater of the same temperature as at the depth where they had been taken. The next morning he found them covered in little white flowers. He took the branches out of the water and the flowers

---

[4]*Linn. Correspondence,* vol. I, p. 226.

[5]Johnston, *op. cit.*

[6]See Anita McConnell, "The Flowers of Coral—Some Unpublished Conflicts from Montepellier and Paris during the Early 18th Century," *Hist. Phil. Life Sci.* 12 (1990), pp. 51-66.

[7]*Ibid.*

vanished again. He put them back and they reappeared. Over the next few months Marsigli experimented with adding different reagents to the water to see what effect they had on the flowers. Excited by this discovery, he evidently cannot have known that the concept of the "flowers" of coral was already familiar and had actually been illustrated in a medical treatise five years earlier. There, the author had thought coral was a marine stone which somehow grew.

In 1724 the son of a French friend of Marsigli's became interested in these coral flowers and called on him. Jean-André Peyssonnel repeated all the older man's experiments together with some of his own. His findings convinced him that what everybody thought of as a flower was actually *"un insecte"* like a little octopus. He wrote a letter to the Académie des Sciences which was never published but whose ideas were later addressed by Réaumur and demolished in the lofty fashion of a grandee confronting the work of a junior. Réaumur declared that "what Peyssonnel had taken for animals living in a habitat of their own creation, analogous to the wax combs of bees, were merely insects infesting the coral stems . . . Réaumur himself believed coral to have a dual nature—a vegetable 'bark' growing on a stony core."[8] Snubbed, Peyssonnel waited until 1751, when he sent a copy of his *Traité du Corail* to the Royal Society in London. This was translated and extracts published in 1752, the year John Ellis addressed the Society with his own theory of corals being "ramified animals." Peyssonnel had anticipated him by thirty years.

The idea that corals were not colonial invertebrates but a strange hybrid species lingered for a long time, and even the authoritative George Johnston a century later dissented with this view of sponges because they lacked polyps. He asserted they were plants and quoted in his support a certain Professor Owen writing in the *Lancet:* "If a line could be drawn between the animal and vegetable kingdoms, the sponges should be placed upon the vegetable side of that line. Locomotion could be no proof of animality; for it was well known that the sporules of some cryptogamic plants possessed very perfectly the power of motion."[9] Even in the twentieth century some people still find the living—let alone animal—nature of corals hard to accept. Vice-Admiral

[8] *Ibid.*
[9] *Lancet* 871, p. 225.

Boyle Somerville records an instance when crew members of HMS *Penguin*, ordered to take soundings in the Kermadec Deep near Tonga, became obsessed with the beauty of some reef formations and went to great lengths to snap off and bring back several large chunks as souvenirs. These were winched aboard the ship under the indulgent eye of the skipper (who had actually been in the old *Challenger* in the 1870s) and young Lt. Somerville. There is no indication that either of these men knew any better, and they seemed as surprised and disillusioned as their crewmen when after twenty-four hours in the ship's hold the corals sent a stench of putrefaction throughout the vessel.[10] No doubt even today tourists by the thousand make the same error.

Susan Schlee suggests disarmingly that part of the reason why so many nineteenth-century oceanographers were fascinated by corals was that most of them came from cold, wet northern climates and coral reefs are to be found in balmy tropical and subtropical latitudes.[11] Once it was agreed that corals were animals there was a new problem to be addressed, itself a promising field of research. This was, how could atolls have grown from the deep seabed when the reef-building polyps could not survive below about 200 feet? There was a suggestion that maybe they built on the lip around the craters of undersea volcanoes, but that left geology with the task of explaining how there could be so many volcanoes of nearly identical height, all of them conveniently reaching to within 100 feet of the sea's surface.

There are three main kinds of coral reef: barrier, fringing and atoll. A barrier reef generally follows the coastline, often at a considerable distance, converting the sea between it and the shore to a shallow lagoon. A fringing reef is more like an extension of the land, and in place of a lagoon tends to have flats which may become partly exposed at low tide. An atoll resembles a roughly circular barrier reef, sketchily enclosing a lagoon. Even before the *Beagle* visited the only coral atoll of its voyage Darwin had evolved an elegant theory to explain how a coral atoll might be formed. He thought the process he described would apply equally well to barrier and fringing reefs, since although the disposition of nearby land might be different, the principle had to be the same because of the constraints governing the life cycle of coral polyps. He took for

[10]Boyle Somerville, *The Chart-Makers* (1928).
[11]See Susan Schlee, *A History of Oceanography* (1975).

granted the truth of three assumptions: that corals needed warm shallow water in order to flourish, that parts of the Earth's crust were sinking at a very slow rate, and that corals could grow upward at least as fast as their substrate sank. The idea must have come to him as a result of finding marine fossils in the high Andes, of realizing how portions of the Earth's crust had in the past undergone considerable vertical movement and knowing there was no reason to think this process had stopped. This by itself was a radical idea at the time, as we know from the controversy Lyell's geological theories had provoked.

Darwin proposed a volcanic island poking up out of the sea to whose shallow inshore waters coral larvae drifted, forming colonies. In time these colonies merged and became a fringing reef more or less surrounding the island. The island itself, meanwhile, was sinking and obliged the corals to build upward so as to remain in shallow water. Since the island was cone shaped it shrunk as it sank, producing an ever-widening lagoon around itself. The innermost corals, now cut off from the constant supply of nutrients in the open sea, grew no more, so that when the peak of the island finally disappeared it left a lagoon roughly surrounded by a ring of coral whose outer edge might extend straight downward to a great depth as successive generations of polyps had built on their defunct predecessors. In this way the coral ring of an atoll marked the original outline of a vanished island.

This neat theory of Darwin's was contested, notably by the Scottish oceanographer John Murray. After the sudden death in 1882 of Wyville Thomson, who had led the *Challenger* expedition in the 1870s, Murray took over the job of publishing the expedition's immense report. He had been on the *Challenger* himself and was very struck by the sediment samples which revealed far greater precipitation than he had imagined. He now thought Darwin's theory too complicated and too dependent on crustal movement. Instead, he proposed that the calcareous remains of plankton, falling on undersea mountains in a steady drizzle over the millennia, built up a layer of compacted sediment reaching to within 200 feet of the surface, at which point the coral larvae started their colonies. In order to explain the atoll's characteristic shape Murray somewhat weakly suggested that this was simply the normal pattern of growth for coral colonies. Trees grew into tree shapes, corals into atolls.

In order to settle the dispute, scientists visited the Great Barrier Reef and the Maldives (where the world "atoll" came from) at about the turn of the century, but the results of their researches were inconclusive. It was obviously going to be necessary to bore straight down into a reef and take core samples. If these turned out to consist of dead coral Darwin would be vindicated; if it was sediment, Murray. HMS *Penguin* (without Boyle Somerville on board) sank some holes and the findings seemed to support Darwin, but they were challenged by champions of Murray's theory who claimed the results were false because coral fragments had fallen down the borehole as it was being dug.

The problem was not finally solved until after the defeat of Japan in 1945. Suddenly the United States found itself in possession of hundreds of scattered coral islands and atolls which had formerly belonged to the Japanese Mandate. Needing somewhere isolated to test nuclear weapons, the US searched its new Trust Territories and hit upon the Marshall Islands. At that point a serious investigation was made of the underlying geology, in the course of which scientists took core samples from deep borings. In Eniwetok and Bikini atolls they went down to over 4,000 feet and found it to be solid coral all the way to the basaltic bedrock, interspersed with strata containing the fossils of land snails and pollen which showed that the island's generally downward progress had been interrupted from time to time by violent upheavals which had brought it back above sea level for a few millennia. Darwin's theory was essentially, if belatedly, proved correct.

Since then it has become clear that many reefs (including the Great Barrier) are formed in ways rather more complicated than Darwin's simple schema. In 1919 the American geologist Reginald Daly proposed that glaciation during ice ages would have had just as much effect on corals in terms of their growth and erosion as geological sinking. Current anxiety over global warming and the projected rise in sea levels has revived Daly's ideas, and in places like the Maldives it should soon become apparent to what extent changes in sea level mediate the formation of coral reefs.

~

The centuries-long dispute about the nature of coral is rendered neither obsolete nor irrelevant by modern science. True, there is no longer any doubt about the organisms responsible for reef building and—broadly—how they do it. At levels of biochemical detail, however, there is still much to learn. Most reef-building corals, for example, exist in symbiosis with microscopic algae. A single coral polyp looks very like a miniature anemone, its close relative. It has rings of stinging tentacles surrounding a mouth, all of which is able to contract defensively into its stalk. Living within its tissues are the algae which among other things perform photosynthesis and fix nutrients for their host. More than that, the algae enable the polyp to secrete stone. This is a most remarkable attribute and brings a certain accuracy to the old name "madrepore." Since algae able to photosynthesize are considered plants, there arises the peculiar arrangement whereby a plant and an animal combine to produce a mineral—in this case pure limestone. The chemistry by which this is done is not yet entirely understood. There is great complexity in the way these symbionts interact both with each other and with the nutrients in the seawater, with varying temperature, degrees of salinity (fresh water is fatal to corals, which is why fringing reefs are always broken at river mouths), which currents and with light. Not the least striking part of a coral reef's equivocation, therefore, is that it imprisons at its heart a gigantic plant, while on its surface and slopes there may be few marine plants visible since herbivores such as surgeon fish and sea urchins constantly graze seaweeds back to their roots.

The swimmer among reefs likes to know such things, likes to shine flashlights into cracks and crevices, takes pleasure in seeing a tubeworm sense his presence (smell? sound? currents?) and retract in a flash; takes pleasure also in knowing the worm ate its tunnel into the coral limestone by secreting acid. It is not only in details that we experience reefs, though, any more than we experience a forest by examining leaves and marveling at processes of gas exchange. Both reefs and forests may be studied closely but are experienced as environments. So viewed, coral reefs are true borderlands, abounding in all sorts of ambiguity. Many of these ambiguities are set up by the classificatory systems which have been used to make sense of phenomena that refuse their assigned niches. The swimmer who daily goes down among corals to watch and listen soon becomes aware of something in this rich profusion which corresponds to the so-called "dark matter" postulated by astronomers to account for there not being enough visible matter in the universe to satisfy theory. (In this sense "dark matter" can be viewed as the astronomical equivalent of the vanishing islands of the North Atlantic.) The "dark matter" in reefs is the subject of the second part of this chapter; it is enough for now to point out that any form of classification, merely by drawing an imaginary border between two groups of objects, spurts into sudden being a third space as real as the counterbalanced pole which marks the frontier between two countries but is not in either one. Merely to propose that A is not B automatically brings a third coordinate into play, the offspring of wherever it is one stands in order to dispense categories, and which partakes of both A and B. All liminalities belong entirely to the mind, and we are perverse if we expect the objective world to keep to our categories.

At the moment the whole corpus of Linnaean taxonomy is in dispute as radically new ways of grouping creatures are proposed. The way we view the natural world is unconsciously influenced by our received picture of a stylized evolutionary tree which over the last 200 years has become steadily more ramose and bushy as efforts are made to graft fresh discoveries onto the existing trunk and limbs. Changing this schema would have an interesting effect. There is no serious question that some form of Darwinism is the best model we have of how things evolve, but there is much doubt about traditional notions of the relationships between them. Modern taxonomy has to confront the possibility that the hypothesis of "Mitochondrial Eve," the putative mother of the entire human race, may eventually become elevated to a theory by advances in genetics. The category of *species*, the lowest classification of all, is under constant pressure not only from modern fieldwork but from paleontological evidence like that of the Burgess Shale.[12] Maybe coral polyps will one day find themselves in an entirely new category (or back in an old one) of plant-animals, now based on something like ways of dealing with nitrogen.

Vladimir Nabokov, who was once simultaneously the curator of Lepidoptera at Harvard and professor of literature at Cornell, said of writing that "the precision of the artist should accompany the passion of the scientist." This is an admirable recipe for swimming among reefs, where details must be avidly noticed but never seen in isolation. Modern entomology no longer distinguishes between moths and butterflies (Nabokov's particular interest), the classification having been deemed too arbitrary to be meaningful. All those Victorian clergymen, earnestly sorting knobbed antennae from feathered antennae and mildy put out to discover day-flying moths, were making distinctions which have

---

[12]See Stephen Jay Gould, *Wonderful Life* (1990).

turned out not to be taxonomically useful after all; except that no discrimination is ever wasted in either science or art, since all interest derives ultimately from the ability to spot difference.

~

A reef project (how full the world is of reef projects!) was under way a few years ago off the coast of Kuyo Island. At the time I was living on an islet in the same group. On a visit to town I met Jim Parkes, who had until recently been a Peace Corps volunteer, had since gone away and returned with a technical qualification and was now seconded to the Bureau of Fisheries. He took me some miles to a dismal scene. Offshore was a typical fringing reef which at low tide became a flat, extending maybe 200 meters to a white line of surf. Ordinarily it would have been an expanse of puddles and small pools at low tide, dotted with boulders and slimy with livercolored blobs of this and that. One would have expected to see groups of villagers, mainly children, with tins and plastic bags, going through the tidal pools for edibles: shells, baby octopus, blennies and gobies. Instead, the reef was a destruction site. A new causeway of blinding white coral chips led out to the middle, where a yellow excavator stood up to its axles in seawater, fanged bucket plunging and rising, disgorging a rubble of water and corals into the back of a truck.

"We've got a problem here," said Jim. Up to then I had not especially liked him but was now won over by this unsuspected laconic talent, for the scene was one of ruin made ironic by its backdrop of a tour operator's tropical idyll. A long white beach stretched in a curve backed by coco palms, all done in the three primary colors of holiday brochures (blue sea, white sand, green fronds) whose baldness in print never hints at the possibility of considerable subtlety. "What we've got is a company mining coral for construction and a fishing community without any fish. And a mayor up for reelection."

The coral had been mined here for some years before the heavy machinery came. Men with crowbars had levered out chunks which buffalo hauled away on crude sleds. Coral could be crushed and spread as a gravel bed for roads, or roughly dressed could be used for building in place of the ubiquitous hollow blocks. "Labor's cheap, cement costs," as Jim remarked. "Now, the gentleman who owns the rig and the truck is a local who sort of inherited them from a government project to build a road around the island. That failed from lack of funds and equipment, largely because they were all embezzled or stolen in the first month. I've got this congressman friend in Manila who has his own reasons for not wanting the beaches of this island ruined. . . . No, you don't want to know about it, believe me. . . . So although he doesn't realize it yet, within a week or two this guy with the 'dozer's going to get zapped by Presidential Decree number twenty-eight thousand three hundred and eleventy-six which forbids the destruction of the foreshore owing to the fact that it's national property. Problem is, what to do with the remains."

About half a mile of immediate reef had been destroyed, the topmost four feet having been stripped out. One consequence was that the corals on the surviving outer slope were visibly suffering since the outgoing tide was now bringing with it none of the nutrients and planktonic matter generated by corals inshore. Jim and I swam up the reef and he showed me a further effect of the excavations. There were powerful currents on this coast, in some places running almost parallel to it. Lowering this reef flat by a mere four feet had caused a neighboring bay to be scoured of its sand, which the current was dumping over adjoining reefs, smothering them. Local fishermen had been driven onto reefs farther along the coast, leading to friction with those already there, to overfishing and the increasing use of dynamite in an attempt to make up for smaller catches. This in turn destroyed corals, which led . . .

"You get the picture," said Jim. "It's bad wherever it happens but they're doing this sort of thing even in the Maldives, which has to be cutting their own throats. The highest point in that country's about two meters above sea level, and since it's all coral there's nothing else to use for building. Probably the best thing there would be to earmark an entire atoll for excavation and ban it everywhere else. As it is, when you remove coral piecemeal, like here, it's hard to see the end of the knock-on effect. Last week a fisherman was killed about five miles down the coast. Dynamite. Got the fuse wrong, took off his arm and half his head. OK, you can say that was his fault, it was his choice to go fooling about with gin bottles packed tight with fertilizer. On the other hand you could say he might never have done it if it hadn't been for this thieving dickhead here."

Our swim had revealed something else about the surviving corals on the seaward slope, which was that they were showing the first signs of bleaching. This is a condition which happens when the polyps eject their symbiotic algae. Since it is the algae which give polyps their color (true corals produce pure limestone, which is white), the corals blanch, the polyps becoming no more than a milky slime on their surface. It is not yet known why bleaching occurs, only that it is probably a sign of stress. There may be no unitary cause. The polyps can recover spontaneously or die, in which case the corals often turn green as their skeletons are colonized by ordinary algae. It is unsettling to swim along a reef slope and see among the profusion of coral species the bone-white beginnings of blight, a marine leprosy mysterious and as yet patchy. It conveys only that something is wrong and likely to become worse, that everything is more poised on an edge than one knows.

A few days later I was in the middle of the small island, inside the belt of coconuts and among the paddies and fields of peanuts. Behind the last plantations rose an abrupt low cliff of dark gray rock from which grew creepers and swags of greenery, an escarpment topped with uneven patches of forest. It was evidently from this miniature lost world that each evening the great fruit-eating bats flew out, many of them spanning more than a meter, to descend on the cultivated coastal belt. There they fed on papayas and drank the palm sap being tapped to make toddy. The cliff was no more than 100 feet high, generally less, and was made of fossil coral. Everywhere were the regular patterns of polyp colonies, the dimplings, pores and stars of tiny stone-secreting animals whose labor had built entire islands as well as the Great Barrier Reef, the biggest single structure on Earth. At some time in the past the flat top of this island where I stood among the tamarinds and cashews had lain beneath a shallow sea rippled with sunlight and thronged with blue devils and fuseliers like chips of sapphire. Having sunk and then risen it was presumably now sinking once more. It was difficult to make sense of such geological eras because the inconceivable slabs of time could briefly be prised apart and wedged open by rudimentary arithmetic. If corals could grow at the rate of an inch a year, which under optimum conditions they can, then this entire escarpment might only have taken 1,200 years to lay down—say from the time of *Beowulf* to the present. A mere nothing, really.

Why, then, could it not be dismantled again? If people needed building materials it made more sense to break up dead coral deposits than to destroy live reefs supporting a complex local ecology which included many hundreds of villagers. I put this point to Jim, but he said fossil coral was far harder than recently formed limestone. The outside might appear friable, but that was just weathering; inside it was like granite. I did not know if this was true. At any rate the idea came to nothing and one was left with the sense of a gross imbalance. It seemed abysmally crude to smash up a construction of microscopic architecture in order to build other structures of a coarseness grotesque in comparison. What, then, was Jim's "project" if not merely to stop the mining of coral?

"Reestablishing fish, naturally," he said. "Or unnaturally."

The best thing to do was simply to leave the hacked substrate alone for a century or two, by which time and with any luck new corals would have established themselves. Sure, said Jim, the best thing so long as you had an alternative source of food. Hadn't I got the idea *yet?* This business wasn't particularly to do with corals, it was to do with fish, and eventually with votes. If I tried thinking of the fishermen as farmers, then these reefs were their fields. Plow up a field without resowing it and you'd get nothing but a crop of weeds.

I supposed so. The truth is, remedial work bores me. If one is a fatalist one believes that once something needs to be restored, to be caught and fought for on the edge of extinction, then in a sense it is already gone, has already been lost in the form that had meaning. It is a pleasing irony that any revivalism, like life support, always suggests demise and not survival. Why should this be? And does it mean that someone who recovers after terrible injuries and weeks of intensive care is no longer authentic? At any rate, Jim's solution was to create an artificial reef out of old truck tires. He was going to take delivery any day now of several hundred tons and dump them all over the ruined reef. It was, he said, an experiment which would be watched closely in many parts of the world. If it worked it would solve two major problems at once, since up to now nobody had found any use for old tires except maybe as fenders on wharves. Tires would make terrific habitats for reef fish because if dumped higgledy-piggledy they formed an elaborate honeycomb of holes and crannies for shelter. After a time they would become encrusted with marine growths and maybe eventually even corals would take root and start building. In the meantime they would be colonized by fish.

By now I had grown quite fond of him despite his stainless-steel faith in remedies and projects and even—quite possibly—in progress, too. But away from him I found myself caring only slightly that well-meaning, busy people like him should succeed in their tasks. I knew all about the fishermen but saw no reason for not holding contradictory views at one and the same time. It is not possible to balance an equation between aesthetics and somebody's livelihood. They have nothing to say to each other. As far as I was concerned the dumping of a shipload of old truck tires into a man-made gap in a fringing reef was adding insult to injury. If the equation were to be solved equitably, it would entail the agents of the damage paying its victims compensation for as long as it took the corals and fish to reestablish themselves. Jim said I was a driveling idealist, flitting about the world like a disdainful butterfly irresistibly attracted to decay, content merely to feed off it rather than do something about it. I warmed to him still further, saying he was an excellent judge of character. Butterflies do not believe in the efficacy of *doing*, but they definitely are fond of flitting and feeding and driveling. We bought each other a series of drinks until the mayor, who would be mayor again, came in and began a lengthy discussion of *arrastre*, or lighterage fees, for unloading hundreds of tons of perished rubber.

2. SENSING THE OBLIQUE

Just as law courts in certain hot countries provide a habitat for diverse species of people, so does a reef supply the architecture and ambience for all manner of marine predators and prey, hangers-on, refugees and indigent waiters on scraps. It has its diurnal and nocturnal rhythms but the activity never entirely stops. After a strange twenty-minute hiatus, night creatures follow day creatures through the self-same corridors and chambers and are themselves distinct; but unseen and in no particular place long, long processes continue imperceptibly. Even in the depths of night, scattered lights wink in the windows of these unsleeping cities.

As a ramshackle social nexus a tropical law court is awash with voices. The sounds seem almost generative, as if every voice secreted a molecule of stone and buildings grew out of pure speech. Instead of some general administrative order, it is the uproar itself which supports and enlarges the monumental fabric. Around its walls and beneath the shade of banyan trees sit men at tables with ancient typewriters, patiently transcribing voices from the throng: aggrieved, supplicatory, self-righteous, wheedling. For every sentence typed, a thousand ascend like smoke past the birds in the banyans. Hot knots of waiting petitioners discover each other and put their cases over and over again, ever more articulate, ever more impassioned, scattering in the sunlight gems of rhetoric whose brilliance goes unrecorded. Squeezing through the press of people the tea boys come and go with battered tin trays and dirty glasses with half an inch of undissolved sugar in them. Wholly unconcerned with legal wranglings, they form a noisy subcommunity of their own, their cries and whistles joining the birds'. Everywhere is a prodigality of speech and gesture and smell, and from it all a distillation leaks out in a steady trickle of files containing depositions typed shakily on yellowed paper.

An old man wearing spectacles mended with fuse wire gathers them up every so often and takes them inside the great building, through a back entrance and gloomy passageways, climbing flights of worn steps until, having knocked on a door, he enters a remarkable room. It is not particularly big, but the open windows are spacious and it is the light coming through them which gives the room its quality. This must be the back of the building, facing away from the hubbub downstairs which had seemed to besiege it on all sides. The air is quiet. The windows open into a cage of leaves, as if the room were built in the heart of a tree. Through them filters a serene undersea glow pricked with spicules of dazzling sunlight, one stray beam laying the spectrum in a bar of colors across the windowsill. At a bare table before the window sits a scholarly man—a judge, possibly—staring out into the greenery and watching the hop and flitter of small birds among the branches. The old messenger lays the files in silence on the judge's desk and in silence withdraws, closing the door gently behind him. Maybe something of the din below is audible after all, for there is a faint but constant background noise, a soft roaring like a distant sea. The judge sighs, opens the top file and begins to read. It is hard to connect the orderly, formulaic sentences with the whizzing and tumultuous lives which, in some garbled form, they partly express.

One needs to drift in the green undersea light day after day, month after month, maybe for years, until almost bored. Or maybe not bored but blocked. Looking and listening, the conviction grows that something is missing, some dark matter hidden but deduced. It is an absence which to an animal privileging the visual leads to the idea that it must be *located*, if only it could be found. Somewhere behind the next outcrop (vaguely right-angled, like the wing of a somber old palace) must be a forgotten courtyard containing a great assemblage, a tumult, the core and center of this submarine complex. Until it is found one will go on missing the point, ears stuffed with water and eyes straining behind glass, too immured in a scholarly attention to detail and too intent on deciphering easily apprehended messages which purport to tell the whole story. What is needed is a sideways shift, a skidding off into a different position entirely.

Some black art may be needed for dealing with reefs, one I have never discovered. All my experiments with them, while having suggested themselves as serious, ended up as parodic or whimsical. When diving off coral reefs it would be useful to learn how to judge the direction and speed of a nearby boat rather than run the risk of surfacing for air directly in its path. Even after so many years I still cannot do this reliably. Sound underwater becomes omnidirectional, reflected off rocks, off the shelving sea floor, even back down from the surface. Depth, too, makes a difference. The more distant a boat is, the farther down one has to go to hear it. Sometimes a fishing boat will pass directly above. The silver-edged lozenge scoots overhead—black, very sharp and swift—and the whirring disc of its propeller mutes itself for an instant, like an electric fan which is quieter heard edgeways on than from in front or behind. What, then, would be the effect of hearing a sound from a source vertically below instead of overhead?

Having borrowed a cassette player with a loudspeaker, I swathed it in many plastic bags and walked into the sea, its owner watching and torn between consternation and ridicule. I swam down about twenty feet and set the machine on a ledge among corals. The pressure flattened the layers of polyethylene in a secondary diaphragm across the speaker grille and it was possible to believe equally that this might conduct the sound better or else muffle it. I wanted to see what effect music had on fish and other reef creatures, but also how well it would sound underwater. It turned out to be disappointing, ethereal only in so far as it was inherently weird to hang in water twenty feet down and hear Mozart's G-Minor Quintet coming from behind an outcrop of *Acropora*. It was not so much attenuated as muted, the higher frequencies suffering most. With the cassette player on its back the sound, as heard from the surface directly above, was definitely feebler than at the same distance away horizontally. This was probably due to the focusing effects of rocks and the sea floor. As for the creatures, they paid the music no obvious heed, with the exception of a damselfish which braved the ecstatically depressed sounds to dart at the polyethylene. They are highly territorial, and the player was probably in its backyard. Since fish have excellent hearing, it is likely that this dim source of noise was at all the wrong frequencies to be interesting. It certainly was muffled. This maybe had something to do with water pressure deadening the air column inside the speaker and even slightly inhibiting the vibrations of the paper cone itself. Altogether, it was a pathetic substitute for a GLORIA transponder.

On other days I experimented with making high-frequency noises by partially inflating condoms and trying to make them squeak underwater. It was necessary to wash off the silicone lubricant, otherwise my fingers would skid greasily without making a sound. I finally coated my hands with a resin used locally as the basis of incense and tried inflating the bladders to different dimensions to obtain different frequencies. This, too, was a failure. At best I achieved a brief grunting sound, though it was uncertain whether this was generated manually. Anyone who has ever tried swimming underwater holding an inflated condom will appreciate the difficulties involved.

There are fish which are blind and others which have no olfactory organs, but they all have some variety of acoustico-lateralis system. That is, they have "ears" or sound receptors as well as lateral line organs. A fish's lateral line runs from its head along both sides, its course often marked with a pigmented stripe. It is made up of tiny hairlike sensors which respond to changes in the water caused by local movement and currents. (It was probably a similar mechanism which triggered the tubeworm.) In addition to having "ears" some species of fish have the capacity to make sounds, generally in one of two ways. They either drum on the wall of their swim bladder or, like a cricket, stridulate two bony surfaces together. It is presumed that such abilities are used for courting and mating rituals. Species which have some sort of connection between their swim bladder and their inner ear must have exceptionally acute hearing since the air sac would act like a diaphragm and efficiently collect sound. As for the noises made by marine creatures, some are very loud indeed. Apart from the celebrated carrying qualities of cetacean sound, the sharp snapping of a pistol shrimp is enough to make a swimmer jump, while the grunts of toadfish have been known to set off acoustic mines.

Such things are the background to another experiment, one performed so often as to become a habit. This is to swim out to the edge of the reef on a moonless night, head down into the depths and, holding on to a rock near the bottom, simply concentrate on all that can be heard and seen until the air in the lungs runs out. This is the best way I have yet discovered of apprehending a reef and it has become the central ritual of my explorations. To save blundering painfully into corals, stinging hydroids and sea urchins, one needs to take a sealed flashlight, but it must be used sparingly, otherwise the light destroys the eyes' rhodopsin and leaves one blinder and—strangely—deafer as well. Obviously it is better to choose a reef whose layout and fauna are familiar. Knowing what to expect, it is less of a shock when a stingray explodes off the bottom in a cloud of silt, pelting wings flashing quick beats of white underside as it vanishes into the blackness. It is necessary to be alone and it is *necessary* to be apprehensive. When the night is overcast and the wide drench of tropical starlight falls uselessly into an upper bed of soft cloud and nothing below it can be seen, not even the shore: then is the time to take a deep breath and swim down, grains of luminescence streaming back from fingertips, down and down.

It was whale song which mariners heard filtering through their vessels' resonant wooden hulls and which they took for the Sirens' voices, beckoning them to disaster. To have lain in one's bunk at night and heard on the other side of a few inches of oak and copper sheathing those directionless, distanceless cries must have been to feel the chill of utter melancholy and dissolution. Also, to have felt one's nakedness. This is the effect of listening to reef sounds at night, too. It is more than just the nakedness of wearing next to nothing, and it is more than vulnerability. It is the sensation of animal messages passing *through* one as if, being seven tenths water, one's body were transparent.

At first it is too unnerving to permit concentration. After a time, when nothing life-threatening has happened, the rhythm of swimming down, waiting, and coming back up for air becomes soothing. The sea is warm. With the clamp of water over one's ears and the blackness pressing up against one's mask, conditions approximate very slightly to those of controlled sensory deprivation, a disorienting and eventually unhinging technique fashionable in torture and interrogation circles some years ago. But there is no real comparison. There is too much sensation, too much physical effort in holding the breath, in staying down rather than floating up, in seeing and hearing. It is never more than mildly hypnotic for a few minutes (but with a vanishing of time which makes those moments impossible to calculate). Steady in the background is the loud white noise of uncounted crustaceans stridulating with pincers, horny plates, mandibles, who knows what. Very occasionally it stops dead, and in the ensuing silence a chill passes over the body because a million crabs and prawns have all heard something attention-grabbing which one has missed completely. What is it out there? Out here? The frogs in the paddies do it, too. Nightly they crank up their ranarian machine until it is turning over at a constant speed. It goes for an hour, then abruptly stops. A beat or two, then a few brave ones try to turn it over, are joined by others until it catches and settles back into its rackety tick-over. Crickets will do the same.

Out here we are on the edge of something: of drowning, fear, understanding. The huge unseen city itself seems always on the cusp of vanishing, it is so delicate and its true nature so elusive. It is a place whose strangeness is far greater than we can know even as we painstakingly try to identify each snapping shrimp, each grunting fish, the soft concussion—like a cloth being flapped—of a sizable fish taking evasive action somewhere nearby. *Whup*. But then we, too, are stranger than we imagine. We hang here in the depths with granules of cold fire prickling around us as creatures and currents stir dinoflagellates into luminescence. We hang here, inquisitive carbon-based life-forms, knowing that every atom of carbon now in our bodies was once in the interior of a star. For an instant we dissolve, are without form, become nothing but the point at which the three axes plotting this three-dimensional borderland intersect. The three dimensions of a fringing coral reef are as follows. *Horizontally*, it marks a border between sea and land. The flats at low tide baking for hours in sunlight support a variety of marine animals which can survive out of water and resist a wide temperature range. *Vertically*, the seaward cliff of a reef face reaches to within inches of the surface and may plunge 3,000 feet in a precipitous slope, its initial descent marked by a steady change in colors and life-forms. *Obliquely*, a reef exists at other wavelengths than those we can perceive. The hidden courtyard with its tumult, the babble and rhetoric, the colors and unimagined smells, all are tucked away in great bright pockets of the electromagnetic spectrum which are closed off to us, in sounds we cannot hear, in pheromones our nostrils cannot detect. This knowledge makes us ache, sea creatures that we once were, as for a country we have lost on the far side of a frontier we can barely even discern. We are left with our narrow, thickened senses. We are also left resorting to fossil gestures. I find myself opening my mouth underwater, the better to hear. It works on land, but not noticeably beneath the sea, where a proportion of sound reaches one via direct conduction through the bones of the skull. It is the gesture of a creature with a phylogenetic memory, as if some ancestor with a different otopharyngeal arrangement had opened its mouth to let messages in and out, grinding chattily away with a set of bones at the back of its throat. I also track down a dimly surviving fragment of my lost lateral line, maybe. In daylight one day I notice tiny plumes of silt pulsing from a hole at the foot of a coral outcrop about thirty feet down. I do not know what particular worm or shrimp is busily excavating inside but have the sudden urge to feel this minute puffing as well as to watch it. I put my fingertips half an inch away but can detect nothing. Without thinking I maneuver awkwardly so that my head is close to the hole. The puffing stops for a few seconds but then starts again. Gently I move so that I am almost kissing the hole and can just feel the tiny waves of energy from the unseen creature's fins or pedipalps break against my upper lip. It is the underwater version of that unconscious gesture which makes people press their laundry beneath their noses to feel how dry it is: ex-weanlings whose haptic sense for moisture and movement was designed for the breast.

The reef's vertical axis is most vividly revealed in terms of light, ranging from the brilliance of sunlight to the inkiest depths. The sea is both lit and heated by the sun's energy, which is absorbed and scattered from the moment it penetrates the surface. The uppermost meter of the sea effectively absorbs all ultraviolet and infrared, respectively those wavelengths shorter and longer than the visible. Thereafter, seawater absorbs the longer wavelengths first and at about thirty feet down most of the surviving energy is in the blue-green part of the spectrum. From here downward, increasing numbers of reef creatures are colored in various shades of red. With almost no visible red light remaining they look dark or black and in still dimmer waters farther down become almost invisible. In these top ten meters the simplest experiments show

how much light the water absorbs. Slightly dull objects take on fabulous colors as one swims toward the surface with them. Likewise, a blood-colored anemone becomes pale and anemic as one swims away from it.

At a depth of fifty meters only 5 percent of the sun's energy still penetrates. If the water is exceedingly clear, there is enough light for photosynthesis in plants and algae down to 150 meters. Below that, ordinary plant life cannot exist, which is why there are no great prairies of seaweed covering the deep ocean beds.[13] Generally speaking, the 100-meter mark defines the bottom of the euphotic ("well lit") zone, the most productive part of the sea. Coincidentally, it also marks the theoretical limit to unaided diving by creatures of the upper air. Certain free divers have exceeded this by a few meters (see Chapter VII) but it is at around this point that pressure collapses the lungs. Well before that, the strain of keeping them inflated will have begun to rupture small blood vessels. In splendid violation of a supposed boundary auks (razor-bills and guillemots) have sometimes been seen from submersibles at 100 meters, while the bird diving record is easily held by the emperor penguin with a recorded eighteen-minute dive to 265 meters, or 870 feet. In any case this 100-meter zone effectively denotes the end of the reef as an ecosystem. Below this its dead corals become a habitat for scattered sponges and, of course, the sundry twilight animal species.

Underwater photography naturally has to take into account this absorption of light energy by water. Even if there is enough ambient light for filming, color values will be increasingly affected by the progressive filtering out of the warm and comforting wavelengths: first red, then orange, then yellow. "Correct" color values can be restored by carefully calculated artificial lighting. Any approach to the question of human vision underwater leads back to the fact that the narrow waveband in which our eyes operate corresponds very markedly with how sunlight is transmitted in seawater. Quentin Huggett at IOS has wondered whether

[13]This statement wants qualifying. Plant-like structures may occur in the deep and highly specialized ecosystems of vent communities. These grow up around "black smokers" at volcanic sites and are not based on photosynthesis. The bacteria of vent communities, on which populations of tubeworms, huge crabs and oysters and other creatures depend, have metabolisms which use sulfur in place of oxygen. Hence these strange pockets of life have nothing to do with the upper world and its biochemistry. Meanwhile, the greatest depth at which conventional plant life has been found is 269 meters, a clump of maroon algae in exceptionally transparent water off the Bahamas.

this setting of our visual "window" reflects our aquatic origin. In air the bandspread is much wider. Several species of insect see at ultraviolet wavelengths, while the pit viper uses infrared sensing to detect warm prey in the dark. Had *Homo* developed differently he might have "seen" a different world.

It is sad that we cannot smell things underwater, though now and then something lodges in a taste bud or receptor to produce the simulacrum of a smell, a pungent impression located somewhere in the muzzle part of the face, neither precisely smell nor taste. Sad, too, that our hearing is not very acute, and with a limited range from about sixteen cycles per second up to 20,000 cps. A mere cat has three times this range, a bat six. The pitch at which human ears are at their most sensitive is that of "a child's or woman's cry," according to Yi-fu Tuan, who thinks this suggests that our ear is adapted to favor our species's survival rather than hunting.[14] This is most aggravating to those reef haunters who have no wish to listen to children and women crying, either above or below water, nor any interest in the survival of their own species, but who would dearly like to eavesdrop among the courthouses, malls, temples, palaces and suburbs of their passion.

My lame experiments sound foolish. *Are* foolish to scientists like those aboard *Farnella,* who could suggest any amount of gear for increasing my sensing ability. Yet in the early days of submarine warfare the help of people with musical knowledge and perfect pitch was sought in order to classify the sounds made by submerged craft. In World War I the composer and conductor Sir Hamilton Harty was called in by the British Admiralty's Board for Invention and Research to identify the most likely frequency bands of hull and propeller noises, "anticipating by a whole war a similar attempt in America, where the conductor André Kostelanetz was approached for much the same purpose . . ."[15] Ernest Rutherford also took a colleague with perfect pitch out in a small boat as part of the war effort. At a prearranged spot one of the great names in atomic physics took a firm grip of his companion's ankles while this man stuck his head into the Firth of Forth and listened to the engine note of a British submarine. Hauled back into

[14]Yi-fu Tuan, *Topophilia* (1974).
[15]R. V. Jones, *Reflections on Intelligence* (1989).

the dinghy and toweling his head he announced it was a submersible in A-flat and he would recognize it anywhere.[16]

Today with an ultrasound detector I could wiretap the citizens of this submarine city, but it would not be the same. As with medical diagnostic equipment such apparatus can also blunt other responses. Ironically, part of the problem of the tropical reef is its very *visibility*. It famously stands as the icon of marine exoticism: brilliant colors, profusion of species, intricacy of shape and design. To make all this more easily seen by means of scuba gear, or more easily heard with ingenious electronics, is somehow too facile to be serious, except for detailed scientific work. Scuba equipment is, of course, indispensable for any work below forty or fifty feet and down to 200 feet. It is a marvelous invention.[17] However, it has certain disadvantages which are not compatible with traveling alone in remote parts of the world, since it is heavy to lug around, expensive to replace when stolen and limited by the local availability of a reliable compressor, to say nothing of the obligatory diving companion. Apart from wishing to be quite alone in the sea, I dislike scuba gear for two additional reasons. It is awkward to wear, all the time making me conscious of itself, of bits of metal and tube, of tanks as bulky as a growth, of belts and dials and rubber straps and harness. But even more important, I cannot hear with it. Breaths and bubbles rattle and roar in the ears, the very heart drums in the air hose. It all gets in the way as much as it facilitates, and on most occasions it is preferable to sacrifice depth for the immediacy and greater effort of free diving. The immediacy is whatever transcends discomfort and inconvenience, leaving one uncluttered on borrowed time. By making certain things too easy, scuba

---

[16]See *New Scientist* 1768, p. 80.

[17]Scuba gear is a vast improvement on all its predecessors, although the tonic effects of oxygen seem not to be as pronounced nowadays as they used to be. In a report in the New York *Daily Times* (24 August 1854) the writer considers the newly developed diving suits of the day, which were made of leather and rubber and entailed the divers carrying "a box of condensed air." "The condensed air they are forced to breathe, furnishes them a greater quantity of oxygen in a given time, and increases their strength very much for the time being. A diver, at a depth of ninety feet under water, at Portsmouth, England, was known to bend nearly double an iron crowbar in his work, which resisted the strength of four men at the surface." Presumably the proposition here is that if oxygen gives life, a lot of oxygen will give a lot of life.

equipment gives rise to that curious paradox: The more accessible a thing becomes, the harder it can be to see.[18]

Besides, we are not trying to push outward the frontiers of science, but our own. We are content to have identified the calm room with the big windows letting in the blue-green undersea light and to know that whatever we learn is only a part of what goes on in its corridors and undercrofts and courtyards. It is a consoling pleasure to have hung about this cityscape at night and watched the lights winking to codes un-guessed at, to the roar of conversation in a universe off at an angle. It is enough to have clung to the roots of a reef at sunrise and watched the dark gradually bleach from the water. The night shift changes, silence falls. Then the creatures of day emerge and all around the dawn chorus starts, as if invisible behind still-misty water a great banyan were spread-ing its branches.

[18]C.F. Wallace Stevens's poem "The Creations of Sound," in which he chides an imagi-nary writer whose poems "do not make the visible a little hard/To see . . ."

3. SOUVENIRS

Tourist markets laden with trinkets carry their own charge of melancholy, while seeing the *objets'* provenance can be brutally sad. People in remote places labor long and hard to live by this trade. The sacks and bundles pile up in bamboo sheds or beneath a covering of dead palm fronds: the shells and corals dragged from the seabed and rotted out in heaps; the hardwood cigarette boxes stacked in piles; the sharks harvested for flesh and teeth and jaws; the alligators stuffed; the baby porcupine fish inflated and lacquered; the aquarium fish poisoned.

Often it seems the more that people become urbanized, the more they want about them talismans of nature on their walls, their shelves, their key rings. Many souvenirs are marks of pilgrimage, like religious relics, and denote travel. Many of these talismans come from the sea. They are tokens of lineage and are to *Homo* what a family crest is to an aristocrat. The bloodline lives on. Yet perversely, this importing from one universe to another, from water to air, is invariably fatal. Nothing looks as dead as a seashell in suburbia, a piece of coral skeleton or a stuffed fish. The otherness of a shark's tooth, like that of a fossil, begins to ebb the moment it is held in the hand. *Objets trouvés* should be marveled at, then allowed to become *perdus* at once. Only thus can the transient pleasure of crossed trajectories be sustained in the memory.

The tourist's trinkets and the traveler's memorabilia so swiftly decay from prized objects into junk because they are never what they were believed to be. As pieces chipped from nature they have a status oddly close to that of a monument. Just as a monument purports to refer to the past but is always contemporary,[19] so the tourist's relic constantly rewrites his version of its origin or the moment he acquired it. The more it tells him of his former self, the more silent it remains about its own past. A shark's tooth is to the living actuality of the fish which once ate with it as a holiday photograph is to the scene it depicts. A wave washes over us as we hold them but it is not—as we imagine—a wave of cheerful or tearful recall. Such trinkets commemorate a moment not of acquisition but of loss. The tooth, the coral, the hardwood, the fish, were all wrenched from life so we can later discard them as the impedimenta of a previous self. From long experience we know this in advance, remembering the fate of souvenirs from former holidays even as we tell ourselves that this time they are authentic. But the things are already dead when we buy them; and their rattling into the garbage can later is the sound of our own hollow attempt to seize the evervanishing present.

In this way our own deaths prey on the deaths of others. All mementos immediately turn into species of fossils. The intricate tracery of a coral fan, torn last year from a seabed off Barbados and bleached for export, stands as mutely on a mantelpiece as the ribs of any Carboniferous leaf embossed on a piece of coal. Our psychic recognition that last summer might as well be ninety million years ago robes our trinkets in their characteristic sadness. Acts of memory are incapacitating, trapped as we are between *souvenir* and *memento mori*. When the objects have collected enough admiration and dust to force this recognition upon us, we throw them away or donate them convulsively to a museum.

[19]This idea was developed by Mark Cousins in a lecture at the Architectural Association in London in March 1991.

A death at sea

But there is no reef after all. The swimmer pursued the dark shadow in the water, counting his strokes, until he realized it was receding at every stroke. He gave up and swam in what he hoped was the reverse direction until now he thinks he has returned to his original position.

Still no boat. It has made no effort to materialize during his brief absence. Maybe there never was one. Maybe he has been out here in this radiant deep for days, even weeks? Only the rope's empty tugging at his ankle as he swam reminded him of how recently he had been anchored to his precious craft.

Surely the sun is lower than it was? For the first time he considers what it will mean to spend the night out here. He is not afraid of the dark, nor is he afraid of the sea at night. Yet out here the sharks are maybe not so timorous as they are in the shallower waters near a fringing reef. They like the freedom of a good depth beneath their bellies. Since he will be neither kicking nor struggling he has hopes of not attracting their attention. Like certain predatory humans, they are beautifully attuned to the sounds of distress. No; he will hang here quietly, counting the stars or trying to see the lights of land. Perhaps it will be impossible to resist peering into the Stygian, sparkling gulf he treads, even though the sight of a great swirl of luminescence turning suddenly in his direction might well herald his end. No doubt the strike will be shockingly abrupt, but for some reason he is more frightened by the thought of a first inquisitive tugging at the painter hanging beneath him like a bellpull. He does not want to hear his own terror pealing out across an empty ocean. At all costs he will be stoical. Fishermen say shark attacks are painless. One feels great bangings and slammings but no actual pain.

This is stupid. It is still broad daylight and the boat has only just gone. There is no possible way in which it could have drifted more than fifty or a hundred yards. A hundred and fifty at the outside. Well, he will not wait passively in the water until death in one form or another heaves up alongside to gulp him down. He will now try to swim a grid pattern. It will not be easy without any fixable starting point, but nevertheless he will try.

He heads for the sun, fifty strokes, then turns at what he calculates to be ninety degrees and swims ten more; then another and back parallel to his original course. He is pleased to find the surface of the water undulates more than he had thought. This convinces him that the chances of happening on the boat are excellent. No sooner has he thought this than, halfway along the new leg of his search pattern, he spies its prow. There is no question. No piece of water was ever that shape, black and slender and curved like a beak. With a cry he abandons the stupid grid and heads for it. Thank Christ, and about time too. That was close. God, that was close. Never, ever again will he do anything so damned stupidly careless. . . .

There is nothing. No prow, no boat, not even a length of floating timber. Nothing but empty water taking its shape from him. His grid is broken and lost. He will never find his original position now. He turns and turns in despair and dejection, incredulous to think he will not be seen ever again while the bloody boat will probably be wrecked on some inshore reef to gladden a poor beachcomber looking for useful spars and panels of marine plywood.

IV

WRECKS
AND DEATH

1. WRECKS AND DEATH

Many must have had the fantasy of the sea vanishing and leaving its bed open to inspection, though the imaginative leap may be accompanied by uneasiness about the sea's sudden return, catching the fantasist out in the middle, bent over something of interest. (The ominously rising breeze, the horrid realization that the low strip of dark cloud on the horizon is actually an onrushing wall of water two miles high.) Treasure hunters, especially, must all have entertained the fancy of strolling offshore among the wrecks and hulks drying in the sunshine, scattered as far as the eye can reach. The image is likely to be sanitized, based on pictures of abandoned World War II vehicles in the Libyan and Egyptian deserts, of perfectly preserved scout-car chassis up to their axles in sand. The reality would be one of draped and stenching putrefaction.

Wrecks have a particular fascination because they act as foci for so many preoccupations: death, loss, things being hidden and disappearing, things being discovered and reappearing, hoards of wealth. Irony is added in that wrecks and their contents are frequently quite close to their searchers. They may be no more than a few hundred feet away, but the marine universe into which they have passed makes them as inaccessible as if they were miles distant or on the Moon. If one were to classify them, wrecks would conveniently fit into four categories: tombs, time capsules, gold mines and time bombs.

~

On Pier 40 of Honolulu Port, from which the *Farnella* sailed, stood a huge warehouse, practically empty. I learned from the security guard that until recently it had been used as a dump for impounded contraband and, indeed, below the wharf on the side opposite the ship were moored various oceangoing yachts and catamarans, all of which had belonged to drug runners now behind bars. Their craft were waiting to be auctioned off by US Customs. Before that, the warehouse had contained one of the largest collections of aluminum coffins in the world.

Honolulu, and especially Waikiki, strikes many visitors as moronic in its empty-headedness. It is exactly the sort of place disgraced dictators would choose to go into exile, live in a tropical wedding cake of a villa and die protesting their innocence, all of which ex-President Marcos did. But the island of Oahu has long had a grimmer aspect, one which runs reassuringly beneath the sun-and-surf package tourism. Only a few miles along the coast from Waikiki is Pearl Harbor, while the air base of Wheeler Field at the island's center has been associated with death ever since World War II. For nearly forty years it was the stopover for American KIAs from the Philippines, the Marshalls and Solomons, Japan, Korea, Vietnam and the rest of Indochina. Those killed in action in every Pacific and Southeast Asian theater of war passed through Oahu, and it was here that bodies returned by the Vietnamese government came for identification. Hence the coffins. The bulk of them are now stored elsewhere; but 100 yards away from the warehouse, on the far side of rows of dusty new Japanese cars awaiting customs clearance, was a barbed-wire compound and in this were still a few stacks of the silver boxes, slightly corroded from the salt air and with tufts of weeds growing round the bottom.

It is a positive relief for the visitor to Honolulu to be reminded of war, of anything serious and historic to set against a local culture which is so aggressively frivolous. The dignity of the USS *Arizona* Memorial at Pearl Harbor is a profound contrast. The *Arizona,* which had served with the Atlantic Fleet toward the end of World War I, was one of eight battle-ships sunk or damaged in the surprise Japanese air attack on 7 December 1941. When the attack began a few minutes before eight on a Sunday morning, the *Arizona* had aboard 1,447 of her full complement of officers and men. When she sank, 1,177 had died on the ship, including the division commander and the ship's captain. Some 800 are still there. It was decided to turn her into a National War Grave together with a second battleship, the USS *Utah,* which lies half submerged on the other side of Ford Island from the *Arizona* and still has fifty-eight bodies on board.

The Memorial attracts something like 1.5 million visitors a year. What might have degenerated into just another attraction is remarkably som-ber. The effect is achieved by an organization which suggests that visitors have temporarily passed out of the indulgent cajolery of civilian guided tours and into the gaze of military discipline. One is ushered into a cinema by a trooper who introduces a short film with some background information about what was happening in the world in 1941. The film then shows the attack itself, grainy, jumpy, tilting pictures of hectic black-and-white action intercut with modern underwater footage of the ship as she is now: rusty chains and barnacled capstans. At the end, suitably subdued, the group is led out through another door straight onto a wharf and aboard a launch which heads out toward the *Arizona's* resting place. In this way no ordinary visitor can see the ship without first entering its solemn context.

From a distance one sees the white, perforated Memorial building, half bridge and half observation platform, which spans the sunken ship. The guide emphasizes that at no point does this building touch the *Arizona*. (It was clear from the underwater footage that the ship may only be looked at, never touched. No gloved diver's hand had reached out to wipe away rust from a hatchway or algae from a porthole.) If anything, this launch trip intensifies the solemnity, not least because the tourist's prerogative of continuous smoking and eating and drinking is forbidden. Once at the Memorial the group wanders reflectively around, not talking very much. The older they are, the quieter. There are the ship's bell and a marble wall inscribed with all the names of the dead (as always, men may be sent carelessly to their deaths in wartime but their names are meticulously recorded); yet it is the sunken ship herself which commands attention. The Memorial straddles the hulk and on both sides the ship tapers away, its ends marked by distant orange buoys. A few chunks of corroded steel poke up above the surface, notably a great circular drum, the barbette of a gun turret. Otherwise, the *Arizona* remains shadowy, bluish, submerged. Schools of damselfish nose around the coral growths which have taken hold on her decks, looking for plankton. They are familiar black-and-yellow striped sergeant-majors, though I thought I saw another variety as well, *Abudefduf*. The flitting of these creatures between the observers' eyes and the object of their reflection did not have the same effect as of pigeons circling a cenotaph. They were not tokens of a natural world blithely indifferent to human pieties, but drew attention to the medium into which the victims had passed. The skeletons, the events of 1941 and the fish now inhabited the same world, no part of which had been retrieved for the redemption of daylight and the upper air. We who stood looking down through the fish arranged ourselves in the relaxed, slightly unfocused attitudes of those who musingly watch golden carp in a pond: a quite different posture from the stiff upward gaze of someone confronting a monument.

And herein lies the USS *Arizona*'s unique effect. We are accustomed to look downward at gravestones but never downward at public monuments. This sunken battleship is probably the only example of a monument which is viewed from above. The bowed head is at once a gesture of private grief, public respect and national mourning.

On the journey back in the launch our escort told us that the slight oil slick we might notice came from the two or so gallons of fuel oil which still leak daily from the *Arizona*. Legend has it that it will stop seeping on the day the last survivor from the ship is buried. "I guess the ship is weeping," said a fellow traveler on the bench next to me. He was a man in his fifties with a mustache who told me he had come because he had been a boy at the time of Pearl Harbor and remembered it partly for the emotion and partly because no one at home in Connecticut had known where it was. He had not expected to be so moved, he said, but he retained a sound middle-aged asperity, remarking on the irony of our guide's homeward commentary being full of platitudes about peace, friendship and the lessons of war when only the previous day the UN Security Council had passed a resolution which President Bush was interpreting as giving him leave to go to war in the Persian Gulf.

Back on land I visited the nearby submarine museum. Dotted around outside on concrete plinths and pedestals was a display of missiles, all of which looked oddly small and rudimentary. There were very few visitors and the place had a pleasantly mournful air. There was also a large black "Kaiten" class Japanese one-man suicide torpedo. Apparently this was never a successful weapon, proving temperamental and difficult to control. Essentially a huge bomb with a little seat in it, it sat on its cement bed and bled rust from rivet heads. A notice said that one successful "Kaiten" pilot had gone to his death wearing a white bandana and with the urn containing the ashes of his friend killed in training jammed into the cockpit with him.

The centerpiece of this display was still afloat: a submarine moored to the quay, USS *Bowfin*. She last sailed in training in the early 1970s and the interior looked as though it had been kept polished by use as much as for exhibition. She had survived the last war with a distinguished record of "kills." From inside, even moored submarines give a powerful impression of being on the seabed, an effect only partly to do with the way daylight is rationed by a tiny hatch or two. It is one thing to go down into the depths in a bathysphere of one's own free will, like William Beebe, but surely quite another to go into combat in that blind tube crammed with men and machinery. There was in the *Bowfin* an air of menace greater than could be explained merely by the ever-present pressure of the sea beyond the curving steel sides. Other men's fear as well as extremes of discomfort, perhaps. In the engine room with the four huge diesels going for surface running it must have been hot even with the hatches open. When the submarine dived the engines were stopped and she went down under electrical power. In the already hot, confined space the four diesels went on giving up their heat. The engineers were naked but for shorts. Dehydration was a serious problem; men passed out.

The *Bowfin* produced a quite different effect from that of the *Arizona*. Although both ships had been in combat half a century ago, the *Arizona* had felt as if she belonged even further back, to another epoch. In the museum were old photos of cocky *Bowfin* ratings lolling and smoking in port, draped around the very machine gun up on deck against which I had just been leaning. Young in their forties hairstyles, they had none of the remoteness attending the marble names fixed above several tons of bones out in the harbor. For the real subject of Pearl

Harbor is time, no matter how well it has been displaced onto trenchant exhibits. This at once became clear in the matter of gunsights. It was possible not only on *Bowfin* but also in the shore display to look through various gunsights, ranging devices, binoculars, periscopes and so on. Unlike much else these had not been maintained in working order. Since being given over to the public their focusing and other adjustments had become frozen or disconnected. At most there was a speckly view in one eye. Looking through a periscope toward the distant *Arizona* Memorial across the harbor, I half expected to see a grainy, black-and-white clip from the film we had just been shown, as in the M. R. James ghost story about the pair of binoculars through which one could see only violent and desolating scenes from the past.[1] Indeed, I would swear I saw no color through any of those eyepieces.

So the *Arizona* lies out there for all foreseeable time on the far side of a strip of water which is really too thin as insulation. Crossed as it is by a precise succession of launches bearing (among others) parties of faintly triumphant Japanese pretending to be from Taiwan, it scarcely cordons off the present. It all feels far too close, that nearby coastal strip, the freeways, used-car lots, dune buggies and "All the Ribs You Can Eat for $5" joints. That, roaring by in its oblivion, does not feel like a ransomed world but one which has no use for the past in any other form than in national shrines.

It was at Pearl Harbor I first appreciated how, once it swallowed something, the sea washes it over less with water than with time, so whatever it engulfs becomes ancient almost immediately. It has something to do with being shut off from the continuity of vision, but in a way which is more powerful than mere burial on land. It felt as though the 800 skeletons contained by the still-leaking hulk of the *Arizona* had been borne back and out of history until they and the *Titanic*'s victims and those of the *Mary Rose* or any Phoenician galley were coeval.

With its polished metal and ruthlessly closing watertight doors, the *Bowfin* pulled me back to an event in childhood. When I was between eight and ten years old there was a terrifying disaster involving a submarine. That is, what I remember is my own version in which I imagined what it was like to be a sailor trapped on the seabed in a metal coffin,

---

[1] M. R. James, "A View from a Hill." The field glasses were filled with black ichor distilled from a corpse, and to look through them was "seeing through a dead man's eyes."

eyes raised in silence to the curved ceiling in hourly expectation of the first sounds of rescue. I could not remember the submarine's name, where it went down, or even the year.

What I am sure I recalled were the solemn tones of the BBC's news bulletins: the grave, Home Service accents coming through the varnished wooden slats of our old EkCo wireless. Why the submarine had failed to surface was unclear. It was one of "ours," in home waters, lying on the bottom intact but unable to come up and breathe. Memory has stretched the whole affair over many days, during which I invented everything we were not told. I had the impression of maybe a hundred ratings being informed by a level-voiced captain that if everyone kept as still as possible, breathing as slowly and shallowly as they could while all unnecessary heating and power were turned off, the oxygen could be eked out for maybe four or five days, even a week. Certainly long enough for rescue to arrive. Were not their comrades in the Royal Navy (up there in the sunlight) the most intrepid in the world? With the most advanced rescue techniques? Sit tight, the captain said; help is on its way. And— yes—pray, of course. More things are wrought by prayer/Than this world dreams of . . .

So we huddled over the wireless and kept pace with the rescue attempt as it grew ever more protracted and the BBC's voice ever more solemn. There was, we were told, a plentiful supply of "oxygen candles" on board. These could be used at intervals to release fresh quantities of the life-sustaining gas. What these candles really gave off was *time*, of course: the extra minutes and hours fuming up as the men lay side by side in their bunks in the dim glow of emergency lighting, their whole world canted at an angle and with drops of condensation falling from the ceiling. Only a couple of inches of steel held back the press of black water outside. Some wrote letters home, others played quiet games of chess. Everyone was in good spirits, patient. No one wanted to be the first to say that although their prison was achingly cold, it was also becoming unbearably stuffy.

Until finally, after many days, we up there in the bright sunshine were told there was now no further hope for the crew of the lost submarine. The searchers knew where she lay but she was too deep. Or she was at the wrong angle. Or the special cutting equipment was still on its way from Rosyth. And somewhere hidden away in the black and icy depths the captain, his breathing now painfully labored but his voice still level, would hand round to every man on board a little black capsule. The bulletins stopped abruptly and other news displaced the lost submarine. Ignored, it now lay in silence except for an occasional creaking as it stirred gently to a deep current. After another few days (but how long? Weeks, even?) and in the absence of a living hand to turn it off, the emergency lighting faded to a red glow and finally winked out. Only then did utter darkness cover the dead crew lying in orderly fashion in their bunks, with here or there an outflung arm or scattered chess set to betray a final struggle.

This was how, as a haunted preadolescent, I had imagined the drama to which the whole nation was made privy. Even now, maybe forty years later, I can recapture an elusive wraith of the original terror and sadness I carried about at the time, and visiting the *Arizona* and the *Bowfin* jolted my memory still further. An atmosphere at once solemn and filmic inhabits one's contemplations of all sunken tombs, airless but watertight as they might variously be.

I also remember wanting to know the real details. What happened to the bodies? In the absence of oxygen and in the near-zero temperature, how much would they decay? Would the submarine eventually fill with water from the combined seepages of hatchways and torpedo tubes and sprung plates? Before that would the batteries split, leaking acid to react with whatever seawater had pooled in the bilges and release chlorine? And would that in turn arrest any further bacterial activity, just as in tiny concentrations it could sterilize whole swimming pools?

Or maybe they had finally salvaged the submarine so that one morning it resurfaced, streaked with rust and shaggy with weed? Perhaps they had been advised to use caution as they released the steel dogs securing the main hatch and, as the last one was thrown, the heavy slab was hurled back on its hinges while a roar of putrid gases blew a column of rotting papers, naval caps and pocket chess sets high into the air. And once this dreadful pressure cooker had been opened, the first brave men wearing breathing apparatus and carrying flashlights would descend. . . . Was that how it had been?

So many and efficient are the ways of deferring or obliterating curiosity in adulthood that it was not until visiting Pearl Harbor that I realized I still really wanted to know the answers. The nameless demanded to be named. I decided to track down this doomed submarine, to discover how much I and my friends had embroidered. For instance, it was

inconceivable that the BBC would have described a submarine commander having a supply of suicide pills he could dole out when he judged
a crisis hopeless enough, as if his men had been spies. After some
research I narrowed down a handful of possibilities to a sinking which
fitted all the criteria. In April 1951 HMS *Affray* went down in nearly
300 feet of water off the Isle of Wight. Aboard her were seventy-five
ratings and officers. She also, according to a contemporary newspaper
report, carried "a large quantity of oxygen candles."

The *Affray* had sailed from Portsmouth on the evening of 16 April
on a training exercise, part of which was to involve putting four Royal
Marine commandos ashore on the Cornish coast. She was last heard of
at 2116 that evening, diving south of the Isle of Wight. Her commander,
Lt.-Com. John Blackburn, had been ordered to report daily between
0800 and 0900 hrs. At 10:00 A.M. on the seventeenth, having heard
nothing, the radio room at Fort Blockhouse, Portsmouth, alerted the
authorities with the executive "Subsunk" code. An hour later a search
was under way which over the next two days would involve Royal Navy,
US Navy, Belgian and French craft as well as the RAF. It was thought
there was enough oxygen on board *Affray* to support the crew for three
days, barring damage to the system. There was a suit of Davis escape
gear for each man which included breathing apparatus and immersion
suits.

That night other submarines reported asdic contacts and Admiral Sir
Arthur John Power, C.-in-C. Portsmouth, announced "*Affray* has been
located on the bottom thirty-five miles southwest of St. Catherine's
Point in just over thirty fathoms of water." At daybreak an aircraft
dropped small explosive charges to tell the submarine's crew she had
been found and that ships were standing by to pick up anyone who came
to the surface. Nobody came. Several of the searching submarines reported hearing faint, distorted signals and sounds which might have
been made by someone tapping on a hull, but nobody managed to get
a reliable bearing. In the afternoon the asdic room of HMS *Ambush*
picked up the code letters which meant "We are trapped on the bottom." By this time thirty-four ships were taking part in a search which
was becoming desperate.

Next morning hopes were raised again when an RAF Coastal Command aircraft spotted oil and dropped a smoke canister which was
mistaken by another plane for a marker buoy from *Affray*. That evening,

sixty-nine hours after she had dived, the submarine was officially given up for lost.

Because the *Affray* was only one of sixteen "A"-class submarines it was vital that the reason for her sinking be ascertained. For the next eight weeks the search continued under Captain W. O. Shelford. By early May she still had not been found and Shelford was reluctantly driven to take notice of the large number of letters and phone calls being received at Portsmouth from members of the public who claimed to know where she was. He plotted these alleged positions on a chart and found to his surprise that they mostly fell within a small area outside the main search zone.[2] Shelford told Admiral Power, somewhat hesitantly, and a ship was dispatched to investigate. When it arrived at the location it at once obtained a powerful asdic echo; yet as it turned out, it was not the *Affray* at all, nor any other wreck or rock.

Probably every such sinking generates its own aura which profoundly affects not merely the public at large but those involved in the search. There is an account of a strange experience which the wife of a rear admiral at Portsmouth had on the night of 17 April, the first evening of the search for *Affray*.

Quite suddenly, I realized that I was not alone in my room and in the half-light I recognized my visitor. He had been serving as an engineer officer in my husband's ship, a cruiser, at a time when my husband was an engineer-commander, and we had often entertained him in our Channel Islands home.

He approached me and stood still and silent; I was astonished to see him dressed in normal submariner's uniform although I did not recognize this fact until I described his clothing to my husband later. Then he spoke quite clearly and said: 'Tell your husband we are at the north end of the Hurd Deep, nearly seventy miles from the lighthouse at St. Catherine's Point. It happened very suddenly and none of us expected it.' After that the speaker vanished.[3]

The lady had promptly phoned her husband, who said he had no idea this officer had even transferred to the submarine service, still less that he was aboard HMS *Affray*. Since the Hurd Deep was well outside the

---

[2]For this and many other details of the sinking of *Affray* I am indebted to the captain's own account in W. O. Shelford, *Subsunk* (1960).

[3]Quoted in Warren Armstrong, *Sea Phantoms* (1956), and Edwyn Gray, *Few Survived* (1986).

main search area, ships could hardly be diverted on the basis of a ghost story. That, of course, was while there was a chance the crew might still be alive. It was to be some weeks before Captain Shelford gave enough credence to the other seers and clairvoyants to tell his superior.

Affray was eventually found by HMS *Reclaim*, using an underwater TV camera, a new technology's first major success. She was lying in forty-three fathoms of water on the edge of the Hurd Deep with a slight list to port. It was sixty-seven miles from St. Catherine's lighthouse. At first sight she appeared undamaged. All hatches were shut and none of her indicator buoys had been released. Her hydroplanes were set to rise. Then serious damage was found to her snort tube. This was a hollow mast, thirty-five feet tall, through which the diesel engines could breathe, enabling the submarine to run on her surface engines while shallowly submerged. The cameras found it was almost completely snapped off at the base and the hull valve inside was open. The mast was winched up and examined. It was thought at first there might have been a collision, but it was soon found the fracture had been caused by the design not having been strong enough, combined with faulty material. Probably her commander had radioed his last message, dived, and the snort tube fractured on the way down. Even at a depth of only forty feet, water would have poured through the open valve at a rate of three quarters of a ton per second. It would have been impossible to close the valve against such a flow even if somebody had been standing by it. The water would at once have flooded the engine room and caused electrical short circuits followed by explosions, fire and the release of noxious fumes.

So it seems after all that the *Affray* was overwhelmed quickly and without warning, and that by the time the alarm was raised her crew had already been dead several hours. The "signals" heard during the search, the code word, the tapping, all were imagination on the part of anxious young men with headphones clamped to their ears in submarines identical to the one that had vanished. The powerful sonar return from the clairvoyants' recommended "position" was almost certainly from the DSL, or Deep Scattering Layer.[4] As for the rear admiral's wife, her visitant had been completely accurate.

It would just have been possible to salvage *Affray,* given her depth, the currents at that point and the technical capabilities of the day, but

[4]See Chapter V, p. 185.

it would have been expensive and dangerous. "Since the cause of the disaster had been established little was to be gained from such an operation, her scrap value being no more than about $9,000. Salvage was abandoned."[5] In 1990 her wreck was resurveyed. An officer in the Wrecks Department at Taunton Hydrographic Office added that he himself had been aboard one of the submarines searching for *Affray* and the story of messages being picked up by asdic was "quite without foundation." He said the *Affray* had flooded instantly. "It was chaos out there—messages whizzing about—and no doubt some people imagined what they most wanted to hear. It was a very emotional business, of course." As for the "suicide pills": "Absolute poppycock. Never been such a thing. Pure bosh."

Concerning the bodies themselves, my schoolchild's fantasy had presupposed a watertight submarine. It is possible to form a few tentative theories as to what might happen in the circumstance of seventy-five people dying of oxygen starvation. Much would depend on things such as temperature and the fungi in the remaining atmosphere, though it seems likely the air scrubbers would have removed most floating yeasts. From that point of view a submarine is probably a comparatively sterile place. In any case the hypoxia and high carbon-dioxide levels would initially accelerate decomposition due to venous congestion. This would be followed by a slow process of mummification, during which the skin hardens and the gaseous cavern of the stomach contracts. It is likely that adipocere would form, especially in dependent limbs. This is the condition when fat changes to an off-white, wax-like substance smelling slightly rancid or musty. Adipocere itself being preservative, this would prevent further decay in affected portions. Provided a sunken submarine remained watertight a trip down her interior with a flashlight would probably reveal a majority of mummifying bodies, some partly decomposed, a few even skeletonized entirely. Here and there where a head had been subject to adipocere a face might be seen whose features had scarcely changed.

In the case of sudden flooding, as in the *Affray*, things would be entirely different. The first powerful inrush of water would have caused extensive injuries, including rupture of the eardrums by the rapidly increasing pressure. Since any opening, even one only ten inches across

[5] *Brassey's Naval Annual* (1952).

like a snort valve, would give access to marine animals, the stripping of the bodies would begin within hours. In the sea it is generally the lips, eyes and fingers which go first, being most easily seized by creatures with small mouths or pincers. Cod are especially voracious and as soon as the seawater softens the flesh the remainder will be torn off quite rapidly.

~

Except in terms of size, the *Arizona* is no more of a tomb than the *Affray,* but its status is quite different. It is a national shrine because it fell victim at a turning point in US history, on "a date which will live in infamy," in President Roosevelt's words. Part of her power as a symbol comes from being visible but inaccessible. One can touch a mausoleum; a relative might put flowers on an actual grave in Arlington National Cemetery and pass a musing hand over the carved name. But no relatives may touch the *Arizona.* Not even the Memorial from which they gaze down touches her. Everything about the ship has passed out of the realm of the personal. A further part of her power derives from being monumentally a heap of junk. Most war debris is cleared away, especially if it is blocking a harbor. To have left this hulk in the face of expediency or even aesthetics (for, shorn of its symbolic value, there is nothing very beautiful about bits of rusty steel poking up above the water) is a powerfully contrary gesture. It is a solemn act, going against every urge to tidy away, clear up or edit the past. This being the case, it is hard to imagine what might happen to any blithe infidel who took it into his head to don scuba gear and loot the *Arizona.*

The *Affray* is in a different category altogether. She was a peace-time casualty, though still a military craft. She symbolizes little beyond the misfortunes of happenstance. Most of her interest lies in a collective, brief but intense involvement with the race to save her when, in fact, she was already dead. She is a tomb to seventy-five and cannot be tampered with, though without the same taboo as the *Arizona.* In fact, the whole issue of when undersea tombs can and cannot be touched, and to what extent, is complex. In March 1991 three men with a good deal of equipment dived on the wreck of another British submarine, the *E 49.* This had gone down with all hands in 1917 off the Shetland Islands and was first found by local divers in 1988. The men looted one and a half tons of assorted bronze fittings worth £1,500 (about $3,000) and later, before a court in Lerwick, admitted to "theft by finding." It was stated in court that the wreck belonged to the Ministry of Defence and the men pleaded guilty while claiming they did not know the items were part of a submarine. Two of them were lightly fined and the third was "admonished." At no point in the newspaper reports was there any mention of the submarine's being a tomb or a war grave.

On the other hand the *Titanic,* which sank five years before the *E 49,* has recently acquired untouchable status though only once the technological means for touching her had been perfected and not before a few artifacts had been dredged up and shown on French television. Dr. Robert Ballard, who led the 1985 expedition to film the wreck, has always made a point that nothing should be removed from it, to the extent of throwing back a piece of cable snagged and brought to the surface by an unmanned reconnaissance sled. (He has, however, added to the wreck: a bronze plaque from the *Titanic* Historical Society commemorating the ship's 1,522 dead.) It remains to be seen whether others will be as scrupulous. The ship's best defense is her depth, the secrecy of her exact coordinates and the huge expense of all exploration and salvage technology.

The question stands: When does a wreck pass into the gaze of archaeology enough for the numbers of dead on board to be irrelevant to investigation and salvage? Any famous wreck belonging to this century might be thought safe, on the grounds that victims' relatives might object or sue. Yet that cannot be a hard-and-fast rule, since several modern wrecks have been quite officially looted if they were carrying sufficiently valuable cargoes. There must be some unwritten algorithm which balances the number of bodies against bullion or lost art or artifacts, which only takes shape in words when it is already too late.

Military wrecks fall into a separate category, since in Britain, at any rate, they belong to the Ministry of Defence. This is not a claim likely to be relinquished, moreover, since in certain respects the scrap value of wrecks increases as time goes by. Forty years on, the *Affray* would assuredly be worth more than £5,000 ($9,000). World War I craft have an additional value in that their steel dates from the preatomic age and contains no radioisotopes such as of caesium. This is known as "preatomic" or "aged" metal and came into demand as a direct result of the numerous H-bomb tests in the 1960s. These caused atmospheric pollution in the form of short-lived nucleides which have contaminated every forging or casting made anywhere in the world since then. When scientists wanted to be able to measure very low levels of radiation in samples they were testing there seemed to be no way of making a screen to exclude normal background radiation since the steel of the screen was itself contaminated. The solution was to salvage the gun barrels of preatomic warships. Their advantage was that they did not require reforging. The barrels were simply cut into lengths and the scientists could lower their samples into these massive, inert chambers. For many years demand for aged metal has been satisfied by the wrecks of the German fleet scuppered at Scapa Flow, a rich source of uncontaminated steel as well as of copper, brass and nickel alloy armor plate, which has its own brokers. However, the need for preatomic metal has already begun to lessen with the decline of atmospheric nuclear testing and the decay of short-lived isotopes.

There is obviously nothing new in the idea of bringing up valuables from wrecks. Livy mentions it, and in Rhodes there was a law of salvage which apportioned reward according to the depth from which the treasure was retrieved. In water twelve feet deep a diver received a third of its value, at twice that depth a half. Great fortunes were made by men working with very primitive apparatus. In 1667 William Phipps used wooden diving bells to work in water up to sixty feet deep while recovering $375,000 from a galleon sunk off the coast of Spain. For this he received one tenth, while $169,000 went to the Duke of Albemarle and the remainder was distributed to the enterprise's other subscribers. There seems to have been no claim made by the Spanish Ministry of Defense, nor any particular respect paid to the bones of the crew.

~

It will be seen how readily, at certain points, three of the categories of wreck—tombs, time capsules and gold mines—elide into one another. In time almost any wreck becomes valuable. Sheer treasure will generally outweigh scruples. The perfect example of a wreck too recent to be of any real archaeological interest and too old to be thought of as anyone's tomb is that of the SS *Central America*. This sank 200 miles off the coast of South Carolina in a hurricane in 1857. Of its 600 passengers, 423 went down with her; but they did not prevent the cargo of Californian gold from being raised recently. Probably the single richest treasure ever found, it was worth a billion dollars. "Found" in cases like this does not mean the serendipity of just happening to stumble on a fortune. More searching can take place in libraries and record offices than on location.

It may well be that one of the proprieties of salvage hinges on something as banal as whether or not there are any skeletons left. The action of the sea on human bones is referred to in the next chapter; in brief, much depends on salinity and especially on pressure—in other words, depth. Unloading a seemingly deserted hulk would be quite different from looting beneath the knowing, sardonic gaze of skulls. Like the *Central America,* the *Titanic* does not yet have for us the archaeological interest which made the recovery from the Solent of the *Mary Rose* worth undertaking when it was quite clear that, as a warship, she would not have been carrying treasure. Besides, it had been declared in advance that any artifacts found would go into museums and not salerooms. Sooner or later, but not in this century, the *Titanic* will be looted. It has a mystique about it like that surrounding the tombs of ancient Egypt. What fascinates is not the possibility of finding the odd tiara so much as the retrieval of domestic objects impregnated with this mystique. To be able to eat off one of her dinner plates decorated with the shipping line's emblem of a white star inside a red burgee would be strange until it became a commonplace. Eighty and more years on, the seabed endows an ordinary utilitarian object with a reality slightly at one remove. *It ought not to be here, but it is.* Yet part of it is still in the timeless realm into which it disappeared. Tutankhamen's trumpets have been played after 3,300 years of silence, and if their haunting sound comes at us from another world entirely so would that of—say—a trumpet belonging to one of the *Titanic*'s band, which had last played "Abide with Me" on her steeply canting deck.

Lying off a Philippine province there is a wrecked British cargo ship of the mid-nineteenth century from which storms occasionally wash thick English crockery onto the beach. There is probably nothing of much value on board, since the vessel was merely supplying the outposts of sugar companies on Negros Island. It is odd to see old stone beer bottles so far from home and used as oil lamps in the local huts, as it is to sip palm wine out of china mugs with "Staffs" stamped on the bottom beneath the glaze. The sea seldom gives back much of what it takes and when it does, in random and haphazard fashion, the effect can be striking.

Generally, the sea hides and dissolves, and in translating objects from the upper to the lower world it obscures with "an immensity that receives no impress, preserves no memories, and keeps no reckoning of lives."[6] It is into this immensity that aircraft as well as ships disappear. Somewhere under the waves are the last fragments of Amelia Earhart, Amy Johnson, Glenn Miller, Antoine de Saint-Exupéry and countless others; to say nothing of the redoubtable duchess of Bedford, who was last seen heading out alone over the Wash in her De Havilland Moth in 1937, aged eighty-one.

As for time bombs, these are usually objects which people prefer not to think about too closely. They include literal ones, like the munitions ship *Richard Montgomery*, which sank late in World War II in the mouth of the Medway with a full cargo of bombs, shells and high explosives. It is still there, visible and clearly buoyed off Sheerness. It has been estimated that if it ever does explode it could level the town, and local MPs periodically raise the matter in parliament. Consensus expert opinion remains that the explosives have reached such an unstable condition it is as well to leave the ship alone. Time bombs of a different order are the drums of toxic chemicals such as dioxin, which are frequently washed off the decks of cargo vessels during storms, or dumped as part of an effort at "waste disposal." Radioactive waste is similarly still being dumped at sea in reinforced concrete containers. Oil tankers also go down from time to time without all their tanks rupturing, so that somewhere conveniently hidden lie thousands of tons of crude oil waiting to escape. Entropy being what it is, and oil being lighter than water, sooner or later they will.

Finally, there is perhaps a fifth category of wreck, one which combines the characteristics of tomb and time bomb in that for one reason or another people do not wish to investigate further. On the night of 20 December 1987 there was a collision between two ships in the Tablas Strait off Mindoro in the Philippines. The larger of the two vessels was the M/V *Doña Paz*, a passenger ship belonging to Sulpicio Lines. She was grossly overladen, an everyday occurrence in itself but made worse than usual by the approaching holiday. Somewhere off Dumali Point she collided with a lightless and rusty tanker, the M/T *Vector*, owned by Caltex. There was a huge explosion and a fireball which was seen by fishermen forty miles away.

From the *Doña Paz*, twenty-six survived: twenty-five men and a girl

[6]Joseph Conrad, *The Shadow-Line* (1916).

of eighteen. From the *Vector*, two. To this day nobody knows exactly how many died. Three thousand is the official estimate, but it is certain to have been many more. The passenger ship's manifest showed only the legal maximum, of course, but that will not have been a true count even of those families listed, since children under the age of ten are never included as they travel free. Possibly the dissolving bones of 3,500 people lie in over 1,500 feet of water in the Tablas Strait, but nobody wants to be reminded in too much detail of the world's worst-ever peacetime disaster at sea. At the time world leaders, including the pope, expressed "anguish." President Aquino ordered an immediate "all-out probe." Not until November the following year—and only after another accident involving one of its vessels (see Chapter VII)—were the shipping company's operations briefly suspended. The captain of the *Doã Paz*, who did not survive, was thought to have been drunk and playing mah-jongg at the time of the collision. It was said that the *Vector*'s bridge was completely empty, and the charge filed against Caltex was for carrying "a highly dangerous mix of cargo in a grossly inadequate and unseaworthy vessel."[7] Meanwhile, the victims' relatives formed an association to press for proper compensation. Nearly eighteen months later, in May 1989, Sulpicio Lines claimed "86% of the passengers aboard the *Doña Paz* have been paid for at ₱ 30,000 [over a thousand dollars] per victim." That is, 86 percent of those listed on the manifest. The company admitted no liability for the passengers it had carried illegally and who could not now be identified. Also, of course, the affair had incited hundreds of completely fictitious claims. One way and another, it was convenient that the sea should efface the evidence and close forever over the president's "all-out probe."

[7] *Manila Bulletin*, 20 December 1987, *et seq., passim.*

2. A STITCH THROUGH THE NOSTRILS

The common fantasy of the sea's withdrawing or vanishing to reveal the naked ocean floor shows its involvement with loss. From this derives a special melancholy and a power to haunt. Among the sea's attributes are a capacity to conceal, the ability to stand for time and the quality of erasure.

The ocean's capacity to conceal does not extend only to things one might lose overboard, or to the people, ships and cargoes it has always swallowed up. Its deeps may hide the unknown and the monstrous but they can also veil from sight places which have never been lost but for which some are always searching: Atlantis and those far-off worlds whose presence is less a cartographer's fiction than a fact of desire. This is not to disparage the imaginative space they occupy but to wonder if the future identification of a particular lump of seabed with the fabled land, the Happy Isles or Mayda itself might, after the initial euphoria of vindication, not please its supporters as much as they thought and even mean their having to find something else on which to fix their search.

There is a need for the myth of the lost land which so often has utopian or golden significance. Interest in a submarine Atlantis grew as increasing exploration and travel removed areas of dry land which might plausibly still conceal Golden Khersonese, Ophir, El Dorado and the sundry Lost Cities of higher civilizations. (At the turn of the century Krupps of Essen spent half a million dollars looking for El Dorado in the Mato Grosso, and in 1925 Lt.-Col. Percy Fawcett was murdered in the Brazilian jungle by Calapalo Indians while looking for the ruins of Atlantis or one of its sister cities.[8] As for Conan Doyle's *The Lost World* [1912], it is significant that he visualized it as an island: a raised plateau surrounded by a great escarpment of practically unscalable cliffs.)

The loss of the *Titanic* was not a myth, yet both she and the search for her have taken on something of a mythic status, combining semi-archaeology with the qualities of a quest. Since this is a secular age, sacred relics will no longer do as quest objects (the recent demotion of the Shroud of Turin from holy trophy to medieval forgery ought to have dealt the final blow to the sacred object industry). Things swallowed by the sea will do excellently in their place, however, especially if there are TV rights to their finding. Dr. Robert Ballard's search for the *Titanic* was on his own admission obsessive, and such things arouse wide interest. "My lifelong dream was to find this great ship, and during the past thirteen years the quest for her had dominated my life."[9] It had also cost huge sums of money. Yet the ideal thing about this quest was that it could be rationalized by turning it into a research project. In the 1970s the Woods Hole Oceanographic Institution decided to increase the depth range of its deep-sea submersible, *Alvin,* from 6,000 to 13,000 feet. Since 13,000 feet was roughly the depth at which *Titanic* was supposed to lie, searching for it would be the perfect way to test the improved submersible and a new generation of robotic vehicles which could be deployed from *Alvin* on cables and guided by remote control. These in turn would be prototypes of the entirely free-diving robotic

[8]For a scholarly treatment see L. Sprague de Camp, *Lost Continents. The Atlantis Theme in History, Science and Literature* (1970).

[9]Robert D. Ballard, "A Long Last Look at *Titanic,*" *National Geographic,* December 1986.

sleds which are intended eventually to replace manned submersibles altogether. The enterprise was further legitimated by first being put to military use. The Woods Hole research vessel *Knorr,* needing to try out the *Argo/Jason* equipment Ballard had helped develop, made a practice run over the wreck of the US nuclear submarine *Scorpion,* which was lost with all hands in 1968. The wreck was comprehensively filmed though the pictures are still classified and have not been released. Thus exonerated, Ballard could turn his attention to the *Titanic* and later that same year, 1985, did indeed find her.

It is a happy man who can spend other people's money and indulge his own ingenuity to fulfill a lifetime's ambition. The triumph of Heinrich Schliemann when he discovered Troy stood for Freud as the perfect image of happiness, but it also begged questions about what exactly was being satisfied. The problem of Troy was that it had disappeared to the extent that people wondered if it were mythical, or maybe a composite like Homer himself. To have proved it existed and to have stood in its ruins must have been even more exciting than finding Tutankhamen's tomb was for Howard Carter. Troy had been legendary for 3,000 years whereas the young king was a footnote in dynastic history. Carter had only dangled his name in front of his sponsor, Lord Carnarvon, to induce him to let him have a last dig in the Valley of the Kings, which by then had been turned practically upside down in the search for dead pharaohs.

Dr. Ballard's quest was like neither of these, precisely, in that it was not an act of archaeology. Everybody knew the *Titanic*'s fate; it was no legend. Nor was there any mystery about how she had sunk, as in the case of *Affray* and *Scorpion.* Bland reasonableness ("It's down there somewhere and I'm going to find it") is always unsatisfactory as an explanation for a life's dream and thirteen years of searching. When in such cases the search is described as being more important than finding the object, one is entitled to ask what is really being looked for. Such avidness is normally reserved for things of great personal significance that one has lost oneself, and soon gives way to resignation. What private thing, in short, did the *Titanic* stand for? The question must be left unanswered, but it should be asked. What can be said of such ventures is that they seldom stop there.

Dr. Ballard did indeed go on to find the *Bismarck* and, no doubt, much else besides. In the long sequence of searchings and findings, searchings and findings such men undertake, success is a temporary setback, a resting place on a much larger and grander journey to find the one thing that will satisfy a loss which can never be specified. The sea is the perfect place for it, since whatever it hides beneath its dark leagues of surrogate tears it makes timeless. One might announce one was looking for a lost submarine but a thick, wounded shadow dimly glimpsed at the edge of a monitor screen would seem a thing immeasurably ancient in its melancholy, weeping "rustsicles" from its iron plates, and the thrill of discovery would always carry with it the minor guilt of intrusion. It would also be imbued with the knowledge that what one finds never fits the cavity which the search hollows out. "In a way," said Dr. Ballard of the *Titanic*, "I am sad we found her." Such hulks will always feel privately significant to their finder. At first they may appear to be playing a game with him, being deliberately, almost flirtatiously, elusive. Later they become in their stately woe repositories for his own unassuageable loss. So it was that Dr. Ballard identified with the object of his quest to the extent of expressing anger at the teredo worms which had devoured her woodwork. "After years of gluttony the creatures starved and dropped dead at the table. I have no sympathy for them . . ."[10]

In the circumstances it is perhaps not surprising that so much emphasis should be laid on *not touching*. This perpetually lost object cannot be touched because at that instant it will turn into something else: an ordinary ship, an ordinary battleship, which sank, which has a salvage value, which will attract looters. Not to touch leaves it exclusive, an object of vision, in some sense still not wholly found, while heightened as the public myth. The underwater camera records the details, the mystery remains intact. How could one not sympathize with a man who dreads to think his adroitly publicized private quest might merely stimulate greed rather than a proper solemn wonder? It would be as if Sir Percival were to learn that his companions on the Quest only wanted to find the Holy Grail so they could melt it down.[11]

[10] *Ibid.*

[11] In the epigraph Robert Musil chose for his novel *Young Törless* (1906), Maeterlinck wrote:

In some strange way we devalue things as soon as we give utterance to them. We believe we have dived to the uttermost depths of the abyss, and yet when we return to the surface

Given the sea's power to swallow and erase, it is not surprising that it should have been seen as the lair of malignant spirits. The British sailor's personification of these was Davy Jones, to whose locker the drowned were consigned. He is thought by some to derive from "the Devil Jonah." Jonah was thrown overboard as a sacrifice by his shipmates to pacify the storm for which they held him responsible, since he had angered God by disobeying orders to go to Nineveh and remonstrate about its sinfulness. Jonah, preferring a quiet life, had slipped away to Joppa instead, where he found a boat bound for Tarshish at entirely the other end of the Mediterranean. His locker turned out to be the stomach of a whale. The idea of sacrificing somebody to a storm god was no doubt ancient. What is curious is that the practice has lingered for so long, commuted to the passive form of refusing to save the drowning or to touch the drowned. Bryce, the peddler in Sir Walter Scott's *The Pirate,* observed, "To fling a drowning man a plank may be the part of a Christian; but I say, keep hands off him, if ye wad live and thrive free frae his danger." This is not quite the spirit of Grace Darling.[12] Several widely separated cultures have been recorded as sharing the refusal of certain Scots villagers to pick up a drowned body for fear of meeting the same fate. A description of such behavior can be found in a nineteenth-century account of a ramble in Austria. It is true that the incident concerns the River Danube rather than the sea, but the principle is undoubtedly the same:

In Upper Austria I witnessed an unfortunate crew member knocked overboard by the tow-rope. He fell into the waves and eventually drowned, but not

---

the drop of water on our pallid finger-tips no longer resembles the sea from which it came. We think we have discovered a hoard of wonderful treasure-trove, yet when we emerge again into the light of day we see that all we have brought back with us is false stones and chips of glass. But for all this, the treasure goes on glimmering in the darkness, unchanged.

[12]Grace Darling (1815–42) was the daughter of the lighthouse keeper on the Farne Islands, off the Northumberland coast. On 7 September 1838 a steamboat was wrecked in a storm and Grace and her father rowed out and rescued survivors from a rock. Practically overnight she became a heroine; trust funds and awards were showered on her; a circus made her an offer she had no difficulty in refusing. "Applications for locks of hair came in till Grace was in danger of baldness" (DNB, 1975). She died of tuberculosis four years later, unspoiled by fame and unsullied by marriage.

without a long struggle. During this the rest of the crew, or his shipmates, made not the slightest attempt to rescue him. Showing every sign of not wishing to become better acquainted with the nether regions of the Danube, and probably knowing nothing of the local Lorelei, he fought the waves' clutch to his last gasp. Meanwhile, the crew leaned unmoved over the rail and called in monotonous chorus: "Give up, Jim; it's God's will." This went on until their comrade disappeared.[13]

The author goes on to explain that despite his jocular reference to the Lorelei, local superstition was not concerned with legends of water nymphs beckoning sailors to join them in ravishing underwater kingdoms. Indeed, the boatmen seemed fatalistic rather than superstitious. They simply believed that the river required at least one human sacrifice a year, and attributed this directly to an ordinance of heaven.

Such things testify to yet another unfixable boundary, that which the sea mediates between life and death, between air and water. It is not always clear at what point this is finally crossed. For some cultures and at some periods it seems that merely falling into water is enough to show a victim he has been "earmarked," that from then on he belongs elsewhere. In other places and at other times people are heroically snatched from beyond this flexible threshold and laboriously brought back to life. Timid swimmers may experience the boundary acutely, their upper halves in air and light, their lower in water and feeling the pull of black depths. It is a common enough description: "I feel as though I am being dragged down," as though a great magnet on the seabed were tugging at them, or as if they had entered the gravitational field of a private fate. Others, at home in the water, can hardly swim down far enough for their own delight.

Once, on an island in the South China Sea, I was taken to visit a *mangkukulam*, a sorcerer, who lived inland in an isolated hut by a stream. I was told he had captured a sea devil which was responsible for drowning several local fishermen. It was dead now, they said, but I was on no account to touch it because it still had some of its old power and anyone who touched it ran the risk of being dragged down. Only the sorcerer himself was safe because his power was greater and anyway he hardly ever went near the sea. This man, who turned out to be no more

[13]August Ellrich, *Genre-Bilder aus Oestreich* (1833). p. 12.

than thirty, produced a flattish bundle of purple cloth with a faintly ecclesiastical look to it, something like the material Filipinos use to make the stiff little copes and robes for their Santo Niño images. As he unwrapped it my companions edged backward out of the hut into the daylight, peering through the open doorway. Just before lifting off the last fold of cloth the sorcerer muttered a little chant, then exposed a horror I was only half prepared for.

It was grotesque, about twenty inches long and not quite as wide, black, wizened, and with a terrible face. The mouth was a greedy smile, like a child's drawing or the segment of an orange, while the eyes slanted malevolently and shone dull red. It had four limbs like flippers, a miniature manatee's, slightly curled as if it had died trying to grasp something perhaps the size of a swimmer's leg. So imperceptibly can one take in, over the months, a small community's terms of living that it actually looked to me like a devil, just for an instant, before I saw it was the deformed body of a male stingray. The sorcerer had cut it, molded it, *sculpted* it, setting the gill slits with glass beads and creating from the claspers pseudo-limbs so that the underside—which was on display— became a single nightmarish face. It was, in fact, an excellent example of a Jenny Haniver, a class of grotesques modeled from the bodies of real animals so as to resemble some legendary or imaginary creature. They were very popular in Europe in the sixteenth and seventeenth centuries and many a mermaid was formed (like that which fifteen years ago was still on display in a museum in Alexandria) from the doctored body of a dugong with long tresses of bleached horsehair and fake pudendum. It was the first I had seen and I was not surprised at the fear it aroused. Even the sorcerer himself seemed in awe of his own creation as he covered its potent eyes with cloth and wrapped it hurriedly up again.

It is not easy to say precisely what it is about the sea which so swiftly and powerfully puts things on a mortal footing. Sometimes drowning can seem the least of its lethalities. Even people who live by it and work on it all their lives will occasionally glance up and feel their thoughts and preoccupations momentarily effaced, swept by a bleak melancholy which they sense does something at once banal and radical, like resetting the poles of their existence. It is over in a flash; but the seething waters of the middle distance have something wordless and unpitying to say to the dullest of minds. It is as if the ocean, in certain lights and weathers, were the lair not of monsters or malevolent downward-dragging spirits, but of the fat blank which squats beneath all happiness, flicking out its tongue.

It can happen without warning. We might be traveling from one town to another without giving much thought to where we are, beyond being in a car or train. We look up and catch an unsettling glimpse of sea between hills. At once the day changes, becoming particular, much as it does when we hear of the death of a distant acquaintance. Suddenly we are made aware of a horizon, and over it has come a reminder of outlying importances which have always been there but have been temporarily forgotten. Merely knowing that the sea is near gives the landscape a different feel. For the rest of that day the edge of things nudges in toward us.

To most Europeans, at any rate, "the seaside" is no mere littoral, a bald geographical margin where land happens to stop. It is too closely bound up with the past, with summer holidays, once-yearly pursuits, even erotic adventure, to be an indifferent location. We delve in a cupboard in a winter month, looking for something, and come up instead with a pair of shoes from which sand trickles as from a snapped hourglass. At once we are flooded with memory. We almost hear surf break outside the window in waves which have crossed fields and city streets to find us. Summer's gear with its sad charge is an authentically twentieth-century artifact.

Driving to a holiday place on the coast from a town or airport creates an expectancy which is both fulfilled and intensified by seeing the straight line of ocean off at an angle between trees. *Someone lives here all the year round* used to be one's private thought, but only half envious. That straight line is always of another time, of another land where maybe once one had been happy. Else it stands simply for promise, that tireless harbinger of loss. Anyone who travels is reminded: Over just such a horizon is . . . the land where lemons bloom, where corals lie, the El Dorado or Atlantis of the future.

So coastal towns have quite a different feel from those inland, poised forever on the edge of onward travel or of turning back. They exude impermanence, as if everybody there were touched by this crucial indecision. Even their fabric is subject to it, as though the houses themselves knew that sooner or later they would find themselves eroded away or else stranded far inland. The old port of Dunwich, sometime capital and commercial center of East Anglia, began to collapse into the sea in the middle of the eleventh century and now lies beneath the waves. And on the opposite side of the North Sea are Dutch fishing villages miles from the ocean, cut off from it by the building of the Great Dike in 1932. (Tarshish, come to that, is today buried beneath the marshes inland from the mouth of the Guadalquivir River, north of Cádiz. In Jonah's day, and well before that, it had been an entrepôt on a bay which stretched inland almost as far as Seville.) The idea of no abiding city is nowhere more evident than on a coast, where over the centuries the sea writes and rewrites its own margins.

Maybe this is why in Britain so many old people migrate to the seaside to await death. In order that it should not be too obvious they are drawn to the impermanent to gaze at annihilation, the towns are often built to look deceptively solid and are sited in places renowned for their therapeutic qualities. Yet to retire to Bournemouth is to admit to being in transition. There they are, chairs pulled up in long lines on the promenade, passengers on a municipal voyage. Maybe for some it is a return to a deep, inchoate thing which has been slumbering inside them all their lives. It is death which stretches before them to the horizon: a great absolving sheet beneath which they will slip. What is there to see which makes them stare long hours? The constant flux of waters holds something that mesmerizes *Homo,* though whether it speaks of human origins or of individual destiny is unclear. Moving water has in it a fascination both lulling and imperative. Maybe all continuous movement, whether flames in a grate, crowds in a street, trees in a wind or the flicker of a television screen can catch at the mind and set it in introspective motion. Of all such things, only the ocean never moves without an underlying gravity, even on waveless days of sparkle and dance. No weather is inappropriate for a burial at sea.

Death as a voyage is a common trope and the sea invites embarkation whether the dead are literally set adrift in a boat with a few possessions or sewn into a canvas shroud with the last stitch through the gristle between the nostrils and committed to the deep. Some, still on shore leave, may be seen drawn up on the council's benches along a pier. Before life goes out with the tide the waiting ranks of the elderly may yet be quite unmournful, since their conscious intention in migrating to these, their last resorts, was anything but morbid. Rosily remembered childhood treats and holidays at Blackpool, Margate and Skegness awake hopes of dignified rejuvenation. By retiring to the edge of things a lifetime of unfulfilled summer wishes might be made good or truced. It is rationalized by talk of a milder climate and sanctioned by doctor's orders. Yet even the retired mind must know that most of the year is not summer. In the long winter months after dark, when invisible below the esplanade a black sea raps preemptory knuckles on the shingle, it is time at last to go.

~

Part of the pleasurable melancholy of beachcombing comes from speculating about where the objects came from, what they were, how long they took to arrive. Having been in the sea, jetsam, like wrecks, becomes pickled in agelessness. Even bright fragments of plastic give an impression that they might have been adrift for centuries or a week. A knowledge of winds and currents adds a further dimension of interest and I have spent hours like a maritime Holmes, pacing beaches and building up a mental map of how rubbish circulates in a complex archipelago. Once after three days of storm I walked a deserted coast and came upon the shelving mouth of a watercourse bringing floodwaters down from the hills in the far interior. The sea, still fretful, was tumbling in the surf the things it had thrown up together with what the river had brought down. Scattered around were sodden coconuts, splintered palm boles, empty condensed-milk cans, shreds of nylon fishing net, the dull white sole of a training shoe.

I picked this last object out of the scum to look for a trademark which might give a clue to its origin. It was not a training shoe but the sole of a human foot, perhaps half an inch thick and trimmed of the toes. Its pulpy upper side, long since leached of blood and color, was threaded with nematodes. On the underside were the callosities and scars of a life lived barefoot. Although it smelled I sat with it a short while, wondering when and how each scar had been acquired. It was quite broad and, despite sloughing, still deeply lined in the arch. Not a young person's foot. I imagined a middle-aged fisherman caught in the typhoon a few days earlier and presumed sharks had done the rest. It was hard to see why a shark would snip off the toes and leave the remainder or, indeed, how it could so cleanly have severed a flap of meat. But they are strange and beautiful creatures whose acute olfactory sense makes for impetuosity and abrupt switches of attention rather than thoroughness. Probably the victim had not been on his own and the animal had found a surfeit of food. I threw the sole back into the sea and rinsed my hands. Out under the waves would be sleek stomachs and powerful alimentary canals digesting a cigarette lighter, tatters of denim, a pair of spectacles.

3. SEASICKNESS

Some people do not need actually to be in the water to experience the flexible boundary between life and death. For them it is enough to be on the sea in a boat. They are the victims of seasickness and frequently claim, while in the throes of this ailment, that they would far rather be dead. In this they are in august company. Cicero, having fled to sea to escape Mark Antony's sentence of beheading, was so seasick he gave up and returned to Gaeta, preferring execution to the unconsummated death sentence passed on him by the ocean. Given how long human beings have been seafarers, it must be one of the oldest forms of illness to be described, consistent in its symptoms from culture to culture as across the centuries. No doubt its cause was occasionally attributed to malevolent sea spirits or witchery, but even in Antiquity people were quite capable of being rational about it. Plutarch was curious as to why it occurred only on the sea and not on rivers. He blamed the smell of the sea and the apprehensiveness of the sufferer, perceiving the psychological component which seems to play a large part in the condition. Apparently he never made the connection between seasickness and motion.

There are several seventeenth- and eighteenth-century treatises on seasickness with titles like *Dissertatio de Morbo Navigantium*, for it must always have been recognized as a problem for navies as well as for ordinary travelers, and therefore worth serious medical attention. One suspects it only attracted more general and popular concern during the last century, when there were enough passengers taking part in the mass emigrations to the New World. By the time tourism proper started, especially with excursions from England to the Continent in the latter half of the century, seasickness was the subject of dozens of booklets and articles. Most were more interested in remedies than in causes, the majority admitting that these were not well understood. Among the theories might be any of those suggested by doctors of the day:

(i) an "afflux of blood" to the spinal cord;
(ii) disorientation caused by the rolling or heaving;
(iii) "depression of the circulation";
(iv) "displacement of the abdominal viscera";
(v) the influence of "changing impressions made upon the vision"; (obviously a fallacy, one writer remarked, since the blind are just as seasick);
(vi) the influence of a "marine miasma" or "miasmatical intoxication";
(vii) "sanguine congestion in the brain, provoked and entertained by the deranged center of gravity";
(viii) "centrifugal force within the blood vessels" produced by the oscillation of the ship.

As for treatment, this generally amounted to heavy sedation. Thomas Dutton, a popular medical author writing in 1891, presumably thought travelers also needed the placebo effect of a bizarre regimen. His "cure" began a fortnight before sailing and consisted of a light diet, a digestive pill at night, a glass of salt water twice a week before breakfast and a four-mile walk daily. Three days before traveling his patients began taking a medicine of ammonium bromide and chloroform. Once aboard ship they were to reduce the bromide as far as possible, avoid ship's food and subsist on strong beef essence, dry biscuits and whiskey or brandy and soda. In addition, Dutton recommended any or all of the following: chloral hydrate (favored ingredient of the Mickey Finn, or "knockout drops"), dilute prussic acid, iodine, amyl nitrite, cocaine in quarter-grain doses, creosote, cerium oxalate, sodium bicarbonate, caffeine, eucalyptus and Nepenthe (a proprietary solution of opium in alcohol, dosage as per laudanum). The amyl nitrite was taken orally, diluted in alcohol. Any sufferer on this regime would be doing well if he was even aware of being on board a ship. Many travelers must have spent entire voyages in deep narcosis. In the meantime, starting a "cure" two weeks before any possible onset of the ailment might be presumed to have the effect of almost guaranteeing seasickness. The sufferer had thoroughly prepared himself to be ill, whether from the sea or the prussic acid. The glasses of salt water are odd. They were no doubt emetic, but they might also have had a homeopathic function, as if small doses of salt might make one immune to the briny. That might also go for the iodine, which was derived from seaweed.

Several Victorian experiments were made in which cabins, restaurants and entire passenger areas of a ship were mounted on gimbals so as to remain steady, but these were not a success. The engineering problems were considerable, the boundaries between the "stable" and the "moving" parts obviously being zones of great danger. Such things reflected the consensus that at the root of seasickness was the ship's motion. This was not quite as banally obvious as it might seem. It had not occurred to Plutarch, after all, and until the mid-nineteenth century conditions on board most ships could induce sickness even if the vessel were tied up in port. The food was foul, the sanitation facilities fouler, and contagious disease easily transmitted in the cramped and generally overcrowded conditions. In the circumstances, any number of acute symptoms might mask or exacerbate those of ordinary seasickness. M. Nelken, in his advanced and sensible book *Sea-Sickness* (1856), added an appendix in which he gave details of the safety regulations made necessary by the flood of emigrants leaving Europe for America. The new regulations were an attempt to "apply a remedy to the gross abuses which have caused such vast numbers of persons to be swept into the grave, during the few short weeks of transit across the ocean." In 1848 Congress finally passed an act which for the first time regulated the amount of space allotted to each passenger as well as their total number. Even before Nelken published his book these new laws had effected a dramatic change and "already showed a great drop in the number of deaths aboard to an average of less than 1 per vessel."

Nelken, like Dutton nearly forty years later, ascribed seasickness to motion, but unlike Dutton did not relate it to other forms of travel and perceive it as a special case of motion sickness. "The same symptoms," wrote Dutton, "are often felt by some people, particularly children, when journeying by train or a vehicle, so we may have train-sickness, carriage-sickness etc." This only needs updating with the addition of car sickness and airsickness. Airsickness used to be a much greater problem than it is now because earlier aircraft, like earlier ships, were smaller and lighter. They had only a limited ability to avoid or fly above bumpy weather. Nowadays, airsickness bags are sordid but touching relics of a bygone age. As such they are similar to those few remaining drinking troughs for horses in towns and cities. Once in a blue moon one might be used, filling onlookers with curiosity and pleasure.

The latest medical thinking about seasickness, according to a naval surgeon in Plymouth, is agreed. It is caused by "a discordant clash of information between the organs of balance and the eyes." This diagnosis is charming, being wholly nineteenth century in its phraseology and vagueness. (It even begs the question of the blind.) Evidently the advances of the last hundred years are all in the field of medication, though many people would vastly prefer cocaine to Dramamine. One is left to wonder why, among other things, a clash of information in the head should provoke sickness in the stomach. Each of our two inner ears contains three semicircular canals which are set at right angles to each other in three planes. The canals are filled with liquid whose movement stimulates receptors in the ampullae at the ends of the canals. The swirling of minute particles of chalk suspended in this fluid generally enables the brain to maintain our balance in relation to a stable exterior world. When that world becomes unstable, medical theory suggests, the brain-as-computer (always a doubtful model for that organ) goes into overload, confused by too many variables and a surfeit of conflicting messages from the inner ears, the eyes, the soles of the feet, and so on. Why some people should be more sensitive to this than others is unclear. It is not the same as wondering why some people are allergic to shellfish, because seasickness is not obviously a matter of pure physiology. Psychology clearly plays a significant role and so does habituation. Even poor sailors can in time acquire a degree of immunity or at least control if they have to. To most sufferers, however, the ailment remains like an advance symptom of death. It would surely be hard for such a person to view the sea at all impartially or shorn of its mortal associations.

The wreck of the *Florida* in Jules Verne's *Twenty Thousand Leagues under the Sea* (1870), perfectly illustrating the enduring belief that any object will reach a level beyond which it cannot sink.

Ever since he failed to find the reef he thought he saw and the boat he was sure he had glimpsed, the lost swimmer has become conscious of the gulf he hangs over. At least the empty but navigable plain which surrounds him horizontally spreads itself beneath the sun's broad eye. Finding his way home again, back to life, will be a matter of simple luck or simple physics. A puff of wind here, an eddy there, and he will be reunited with his boat. If for a moment he were able to raise himself only fifty feet above the water, he would spot it at once and the entire traumatic incident would be at an end.

Beneath him, though, lies a dimension which absolutely refuses to reduce itself to a matter of simple physics. The seabed is roughly 1,000 meters away—perhaps 1,500 if he is farther from land than he thought. A mile of water, in short. The swimmer tries to remember what a mile looks like. The entire length of Oxford Street, Centre Point to Marble Arch, but stood on end. As he contemplates this, something unseen like a gush of sepia roars soundlessly up at him from below, without warning, blotting out the sunlit layer which swathes him. This chill black torrent is overwhelming in its despair. It is as though a microscopic ghost had arisen from every test and skeleton of the uncounted radiolaria and plankton bedded on the bottom and had suddenly joined in a great upward fume. Far, far below, the basalt itself is calling in a language of eons and its empty message echoes up and spreads around him in a freezing, inky pool. This tectonic voice paralyzes him. It mocks all human hope. It is worse than his first panic, worse even than the threat of death.

V

DEEPS AND
THE DARK

1. DEEPS AND THE DARK

Why did Dr. Ballard, on his journeys to photograph the *Titanic*, habitually play classical music during the descent and rock music on the way back up? It is unthinkable that it could have been the other way around. The idea of "the Deep" is so powerful that if we listen to the word as we say it, a shiver may pass through in recognition of all the associations it has jarred into resonance. By comparison, "heaven" is blank and thin, even faintly unserious. "The Deep" is utterly solemn. Tennyson, whose childhood on the Lincolnshire coast left the recurrence of the sea and its imagery in his poems, knew the word's exact weight. Stately, funereal, mysterious, it spoke ultimately of loss: a steep dark bulk, time's liquid correlative which gulps down objects, lives, all that was and will be. Sometimes it is dreamily sinister, as when he layers the ocean horizontally to intensify sheer depth and discern his ageless monster:

> Below the thunders of the upper deep,
> Far, far beneath in the abysmal sea,
> His ancient, dreamless, uninvaded sleep
> The Kraken sleepeth . . .[1]

Elsewhere, he blurs the outline of a lost friend with the geological implacability of death:

> There rolls the deep where grew the tree
> O earth, what changes hast thou seen!
> There where the long street roars, hath been
> The stillness of the central sea.[2]

Tennyson had read *Principles of Geology* and must have been struck by the passage where Lyell describes how "many flourishing inland towns, and a still greater number of ports, now stand where the sea rolled its waves.[3] These two themes—monsters and geology—recur over and over again in the intellectual life of the mid-nineteenth century as the question of "the Deep" was finally tackled by science.

~

Alexander the Great allegedly had himself lowered into the Mediterranean in a glass cage whose door was fastened with rings and chains. He judiciously took food with him, anticipating a long vigil. It was a legendary business, suitable for enhancing the heroic myth, paralleled in the twentieth century by stories such as those told during the Chinese cultural revolution of Mao Tse-tung strolling for an hour on the bottom of the Yangtze. In Alexander's case the things he saw were on a properly heroic scale. He observed a monster fish which took three days and three nights to swim past. Such was the insatiability of his scopic drive that, powerless inside his observation chamber, he nonetheless managed to include in his hero's gaze the subjects of another monarch, uninvited as he was to Neptune's abyssal kingdom. It was an exclusive as well as excluding view:

[1]Alfred, Lord Tennyson, "The Kraken" (1830).
[2]Tennyson, *In Memoriam*, cxxiii (1850).
[3]Charles Lyell, *Principles of Geology*, 4th ed. (1835), vol. I, p. 375.

None of the men who have been before me, and none of those who shall come after me upon the earth shall see the mountains and the seas, and the darkness, and the light which I have seen . . .[4]

The glass cage was all-important. Alexander was not in that humbler but more reliable alternative, a wooden barrel with a glass spyhole. Evidently he felt it important both to see and be seen, to be recognized as Alexander the Great by the creatures of the deep. Maybe this is a characteristic of heroes, for it is also told of him that after death his body was embalmed in honey and, according to his own instructions, exhibited in a glass coffin. This must have afforded an attraction of Leninesque proportions, and it would be interesting to know if he left orders that his eyes remain open beneath the honey the better to survey his awed pilgrims even as they him.

The legend of Alexander's descent makes plain that the depths of the sea is no place for ordinary mortals. It took a hero to confront it on anything like equal terms. Despite salvage activities, until the late eighteenth century the average European's mental image of the sea was literally superficial, of a navigable surface above an abyss. It was a treacherous surface, obviously, being liable to spasms of hostility or the unpredictable appearance of awesome creatures from below; but a seafarer needed to know only about winds, waves and currents, and the intentions of other seafarers. Anything deeper was hidden. Yet the work of early hydrographers, the attempts of Scandinavians to correlate the supply of fish with currents, and the demands of geologists to have their theories confirmed made it inevitable that the barrier of the deep sea would be tackled. It was a barrier in more than just the physical sense, however. There was something in the very concept of the abyss which paralyzed thought. There seems no other way of explaining the long survival of certain fallacies more akin to superstitions, often promoted by scientists themselves in contradiction of their own laws. Two famous examples show this: the notion of the compressibility of seawater, and the temperature at which seawater reaches its greatest density.

By the end of the eighteenth century scientists knew perfectly well that water, unlike air, can scarcely be compressed at all. Even under great pressure the density of water changes little, certainly not enough to alter

---

[4]E. A. Wallis Budge, trans. of the Ethiopic version of pseudo-Callisthenes (1933).

its viscosity by much. To the extent that it does change, temperature is a more important factor than pressure. Yet an extraordinary theory survived this knowledge, lasting well into the present century. It held that as pressure increased with depth, seawater grew more and more solid until a point was reached beyond which a sinking object would sink no farther. Thus, somewhere in the middle regions of the great abyss, there existed "floors" on which objects gathered according to their weight. Cannon, anchors and barrels of nails would sink lower than wooden ships, which in turn would lie beneath drowned sailors, who themselves lay at slightly different levels one from another, depending on their relative stoutness, the clothes they were wearing and, quite possibly, the weight of their sins. This notion was reflected in the old saying "Jack will find his own level." The popular belief was that having reached their level, bodies would forever drift and revolve in timeless suspension. After the *Titanic* disaster in 1912 it was reported that some of the relatives of drowned passengers had expressed dismay at the prospect of their loved ones wandering through the abyss in submarine limbo.

In the mid-nineteenth century this belief was held even by many scientists. Doubt had been voiced that the transatlantic telegraph cable would ever reach the seabed Lieutenant Maury claimed to have mapped. Might it not sink only so far, to lie conveniently above any ridges and crevasses? ("This would be to our benefit," one nervous shareholder in the original Atlantic Telegraph Company wrote to a friend. "Yet surely the conversants' [*sic*] voices, subjected to such uncommon compression, may emerge only as mouselike squeakings?" thereby adding a further misconception of his own. Heady days for speculators, in both senses.) In an otherwise sensible book published in New York in 1844 we find:

Heavy bodies, which will sink rapidly from the surface, do at length apparently cease to descend long before they have reached the bottom; the pressure of the water being such as to cause them to remain at certain depths, varying in proportion to their weights. Thus it is that the plumb line will not act beyond a certain length, and we have no means, of course, of extending our inquiries deeper.[5]

[5]Anon., *The Ocean, A Description of the Wonders and Important Products of the Sea*, p. 17.

This passage embodies a strange and interesting idea which reverses conventional heuristic wisdom, namely, that theory can itself make experimentation impossible. At moments like this it becomes legitimate to think about a psychic barrier to exploring the deep. A similar implication, that there are certain things best left undone and certain places it is wiser to leave untrespassed, is no doubt behind the pseudo-scientific reasons periodically advanced for the impossibility of doing them and warnings of the disaster which must befall any attempt to breach the "natural" limits to human activities. It had been predicted at much the same period that speeds above that of a galloping horse would necessarily kill railway passengers. (In this century, too, ideas have been dreamed up to show how any attempt to travel in space would be doomed. Arthur C. Clarke once quoted a man who had written in the 1950s to inform him there was a barrier separating the outer atmosphere from space proper, an "adamantine membrane" which kept our air in. Anyone having the temerity to force his impudent way through this protection put there for our benefit by A Being Wiser Than Ourselves would risk being sucked at infinite speed into outer darkness, followed by all the Planet's air.) The expression of this particular sentiment may be that of the amiably dotty, but the anxiety behind such misgivings has accompanied all technological advance. Those who worry about infringing the depths of space are hardly distinguishable from those who, a century earlier, believed the depths of the sea could never be explored.

Any object or creature floating on the sea's surface is already supporting with its body the weight of a column of atmosphere tens of miles high, a pressure defined as *one atmosphere*. To a creature accustomed to sea-level pressures, such as most human beings, this is not noticeable since his body will be in equilibrium with atmospheric pressure. The moment he descends below the waves, however, he will carry the additional weight of a column of water which, being so much denser than air, bears very heavily on his lightly pressurized body. Water pressure increases by an entire atmosphere for every ten meters of depth. Although the human body is seven tenths fluid, its pockets of air, its cavities, and the materials and construction of many of its components make it far less dense than water. In order to prevent himself being squeezed to death beyond a certain, quite shallow, depth, a human needs to be protected by an outer casing (a diving suit or a submersible) which can resist this external pressure and permit an enclosed environment at his preferred sea-level pressure of one atmosphere. Clearly, the deeper he goes the stronger this protective cell will need to be. The pressure at the bottom of the Marianas Trench is some 1,170 atmospheres, where each square inch of the seabed, or of a body lying on it, bears a weight of 7.75 tons.

The assumption that water itself could be squeezed "solider" by such pressures, as a human body would, no doubt derived from *Homo*'s habit of seeing the physical world in his own image. This fallacious idea that water could be compressed into an impenetrable layer somewhere "below the thunders of the upper deep" was remarkably tenacious although it never stopped serious attempts to take soundings at ever greater depths. Certain scientists did wonder whether the sounding line might not really be going any lower but simply be piling up in loose coils on this invisible floor like a thread of honey falling onto a plate. Samples of mud brought up were explained as the sediment which had likewise collected there over the course of millennia, presumably building up into a false bottom to the sea. It was the second misconception, however— about the temperature at which seawater reaches its maximum density— that was the more seriously and widely held and had farther-reaching effects on early oceanography. It was more baffling, too, since it was blandly assumed—and could very simply have been disproved—that salt water behaves like fresh and is at its densest at 4°C. Strangest of all, the actual temperature had already been established, as Wyville Thomson later pointed out. "In 1833 it was ascertained that the temperature of seawater at its maximum density is −3.67°C, and even before that it was known that seawater can be colder than the freezing point of fresh water and still remain liquid."[6]

Unfortunately, the reluctance to accept what had already been scientifically determined was compounded by the inadequacy of the instruments of the day, as Sir James Clark Ross unwittingly showed on his Antarctic expedition of 1839–43. From the decks of HMSS *Terror* and *Erebus* (which only two years later were to disappear famously when Sir John Franklin took them to the Arctic to search for the Northwest Passage) thermometers were lowered on sounding lines deep into the South Polar ocean. When pulled up, those that had not imploded under

---

[6]C. Wyville Thomson, *The Depths of the Sea* (1874).

the pressure read 4°C. What Ross did not know was that in those latitudes the water temperature drops about 1°C every 550 fathoms, a decrease which by sheer mischance happens exactly to compensate for the opposite effect of pressure on an unprotected thermometer. In point of fact his own uncle, Sir John Ross, had already found a temperature of − 1.8°C more than twenty years earlier with a thermometer which must have been protected against pressure. This was in 1818, during the expedition to Baffin Bay in HMS *Isabella*. Sir John not only established this temperature at a depth of 1,050 fathoms but he also brought up a beautiful "Medusa's Head" starfish (a basket star) from the same depth, something which was conveniently overlooked in the following decades.

The prevailing view of the abyss was now of a vast body of water at a uniform temperature of 4°C, unmoved by either winds or currents. (Had the temperature been allowed by theory to vary between 4°C and − 3.67°C, slow convection currents would have been set up.) Without movement, scientists reasoned, there could be no circulation of dissolved oxygen and no renewal of any food particles in suspension. This in turn would ensure the abyss was "azoic," or lifeless, since a stagnant body of water under huge pressure, at barely above freezing point and utterly without light, could not conceivably support life.

The word "azoic" was coined by Edward Forbes, who in the 1840s tried to discover where the boundary lay between the upper part of the ocean which would support life and this great "lifeless" zone. Forbes was a Manx naturalist who in 1842 sailed to the Aegean in HMS *Beacon* to study the vertical distribution of marine animals. What he found confirmed his theory to his own satisfaction and he came home saying he considered 300 fathoms to be the absolute limit of animal life in the ocean. In fact, as Margaret Deacon has pointed out, such life is particularly sparse at that depth in the Mediterranean.[7] Forbes also went dredging in the Firth of Forth, taking with him the young Wyville Thomson, who for a long time accepted his mentor's "azoic" theory and thirty years later would write shamefacedly:

We had adopted the current strange misconception with regard to ocean temperature; and it is perhaps scarcely a valid excuse that the fallacy of a

[7]Margaret Deacon, *Scientists and the Sea, 1650–1900* (1971).

universal and constant temperature of 4°C below a certain depth . . . was at the time accepted and taught by nearly all the leading authorities in Physical Geography.[8]

If it caused a scientist shame in the 1870s to recall the cant of his youth, it is not easy to know how to treat an episode which took place in the House of Commons nearly a century later. On 12 April 1961 the MP Hector Hughes asked the Civil Lord of the Admiralty, Ian Orr-Ewing, certain questions about recent "experimental missions" beneath the Arctic icecap by the Royal Navy submarines *Finwhale* and *Amphion*. The Civil Lord would give no details, explaining "It would not be in the national interest." His questioner shifted to the less classified ground of schoolboy physics and the following exchange took place:

Mr. Hughes: Can the hon. Gentleman say why the water under the North Pole does not freeze while the water on the surface of the North Pole does freeze? . . . Is the water under the ice kept warm by the heat generated from the center of the earth?

Mr. Orr-Ewing: In view of the hon. and learned Gentleman's interest in bathing, I can understand his anxiety about where the ice forms. If he studies the physical tables, he will find that the water is most dense at 4 degrees centigrade and rises to the surface when it reaches 0 degrees centigrade and starts to freeze.[9]

The "azoic" theory held into the second half of the nineteenth century. When HMS *Bulldog* resurveyed the transatlantic route for a telegraph cable, soundings were taken down to 2,000 fathoms. When the sounding lines brought up starfish everyone maintained that the creatures must somehow have become entangled as the lines were being pulled through shallower levels. It was only in 1860 that the theory was finally and unequivocally exploded when a section of telegraph cable was fetched up for repair off the coast of Sardinia. The cable had been laid three years earlier in more than 1,000 fathoms of water (i.e., over a mile deep) and it was found that various marine animals were encrusted on it, their anchoring filaments having worked their way into the outermost layer of insulation. It was quite impossible to argue that they had

[8]Thomson, *op. cit.*, pp. 56–7.
[9]*Hansard*, vol. 638, p. 235.

"become entangled"; they had quite evidently grown there. Thus it turned out that the abandoning of the "azoic" theory happened neatly to coincide with the publication of Darwin's theory of evolution.

Practically overnight the almost universal conviction that the deeps were sterile changed to intense speculation that they might actually conceal life-forms as well as mineral wealth. As Wyville Thomson was to observe, "the land of promise for the naturalist . . . was the bottom of the deep sea." It is perhaps hard now to imagine the ferment which the scientific method was causing in the middle of last century. In the 1860s wild speculation became common when the abyss was considered in the new Darwinian light, which made necessary a complete revision of prevailing ideas about the Planet's history, about man's position in the "natural" world and about his relationship to a "creator." Two fields of study, geology in general and the fossil record in particular, had a special bearing on oceanography. In his *Principles of Geology* Charles Lyell had avoided tackling head-on the six-day, Genesis version of creation. Instead, he confined himself to pointing out that the Earth's features could all quite adequately be explained in terms of the simple physical processes which were visibly still shaping it: tension/compression and erosion/ sedimentation. This was elegant and satisfactory. The implications might have raised some eyebrows but few hackles, since it concerned only the inanimate world of petrology. The real furor was to come a quarter- century later when Darwin made man and the animals also subject to an evolutionary process which led to notions of trial and error, sports and dead ends, casual extinctions, uncomfortable family connections and— worst of all—to the logical conclusion that *Homo,* far from having been perfected as Nature's last word, must himself still be evolving. Darwin's theory also made plain the crucial evolutionary role played by environ- ment. Where conditions were (in geological timescales) fickle and changing rapidly, the species that survived were those which best adapted and evolved to keep pace with them. It was this idea which, coinciding with the demise of the "azoic" theory of the deeps, generated speculation as to what kind of creature might have adapted itself to conditions hitherto considered inimical to life. All the factors which until so recently had indicated sterility—absence of light, intense cold and pressure, no movement—now suggested the one place on Earth in which to look for unmodified ancient creatures, "living fossils." (The terrestrial equivalent was the search for the "missing link," a hypothetical extinct creature midway between the anthropoid apes and man. Storybook quests such as Conan Doyle's *The Lost World* grew directly out of the notion that living fossils might yet be found on dry land.)

It was as if Darwin's intellectual leap had caused natural laws to rewrite themselves and the invisible "floors," which until so recently had prevented sounding lines from reaching the deep ocean bed, all collapsed at once—like adamantine membranes—and began letting through an array of plummets, grabs, dredges, corers and other sampling devices. Now dredging expeditions began finding sea animals which resembled fossils. Specimens of a stalked crinoid, *Rhizocrinus lofotensis,* were brought up from the deep off Norway. No known modern coastal species of this sea lily had a stalk. This was followed by all kinds of hitherto unknown varieties of starfish and sponges from the world's oceans and further reinforced the idea of "living fossils" which would presumably be more and more archaic the farther down they lived. Though completely wrong, this notion did at least give oceanography the last impetus it needed to begin the systematic exploration of the deep.

Now that the problem was no longer conceptual, the major difficulty lay in designing equipment for taking deep soundings and samples. Where sounding was concerned it was one thing to pay out a weight on the end of a line over the side of a ship but quite another to know when it had reached the bottom, since even the thinnest sounding wire weighed a lot with 2,000 fathoms deployed. Men became expert at judging when the plummet had stopped, keeping a sensitive finger on the line as it vanished overboard. However it was done, it was a laborious process. Thomson recorded that in 1868 aboard the *Lightning* a "Hydra" sounder was used which, weighted with 336 pounds, took 33-½ minutes to reach 2,435 fathoms off Biscay and two hours and two minutes to heave back up again with a few ounces of gray Atlantic ooze. This system was much modified in detail but little changed until the invention of sonar depth-sounding. Even forty years after Thomson, the young Boyle Somerville aboard the *Penguin* was using a more or less identical process. "Birmingham Wire gauge no. 20, galvanized" was paid out over a nine-inch-diameter wheel from a huge drum holding about 6,000 fathoms. The little wheel was connected to a counter graduated in fathoms. There was a complex system of inertia brakes acting on the big spool, automatically gripping and releasing it according to the ship's motions, thereby maintaining a safe and even tension in the wire. The weight on the end was called a "driver rod," actually an iron tube, and was supplemented by two cone-shaped lumps of cast iron. Despite the brakes on the drum the entire process had to be watched "like a hawk," and in fact they lost 10,000 fathoms (nearly 11½ miles) of wire, two driver rods and two deep-sea thermometers before they were successful. Then, "We were the first to see land that came from a depth below sea-level which was just a little more than the height of snow-topped Everest is above it."[10] The *Penguin*'s skipper had been on the *Challenger* with Thomson and related a story which showed that where certain things were concerned, oceanographers had from the beginning had a sense of the priorities. At first, he remembered, the scientists had attached bottles of beer to their sounder in order to cool them. When they came up again from the icy depths the seals were intact and the corks still in place but the contents found to be "the very best seawater." This method having failed, the beer was set to cool in the samples of ooze dredged up from 2,000 fathoms. This was very cold, about 35°F, and no scientific investigations were made of the sample until the beer had reached its optimum coolness.

Pranks aside, that earlier expedition had marked the high point of nineteenth-century oceanography. At Christmas 1872, HMS *Challenger* had sailed from Portsmouth for what turned out to be a 3½-year voyage. At the time it was the best-equipped (at government expense) scientific expedition ever mounted. Some would argue it remains the greatest of all such voyages of discovery. One of the expedition's specific hopes was to find "living fossils," and the scientists aboard vainly sifted ton after ton of bottom samples in search of wriggling trilobites. What they did find was life in even the deepest parts of the ocean. The "azoic" theory was by now officially dead, of course; yet it lingered on in vestigial form owing to the technical inadequacy of the sampling instruments of the day. That is, the expedition's director, Wyville Thomson, observed there was life at both the top and the bottom of the ocean for the simple reason that it was sustainable there. Even the deepest sediments were colonized by organisms such as worms, echinoderms and omnivorous crustaceans, so it was not surprising that abyssal and even hadal (the deepest of all; that is, over six kilometers) ecologies could also support

[10]Boyle Somerville, *The Chart-Makers* (1928)

highly specialized types of fish. Yet Thomson still felt sure that the ocean's middle layer would turn out to be sterile because it lacked nutrients. Such particles as there were fell straight through it and down to the seabed. His problem lay in proving or disproving this. There was as yet no reliable way of taking samples from these intermediate zones without also catching specimens from the upper layer through which a net had to pass twice.

During this long voyage there came a reminder of another enduring myth, and in sad circumstances. William Stokes, a young sailor aboard *Challenger,* was killed in an accident on deck. On the day of his burial at sea a delegation of his shipmates approached Thomson and inquired anxiously whether their friend's body, when suitably weighted, would truly reach the bottom or, as tradition had long maintained, would float at some indeterminate depth. Thomson was able to reassure them that his remains would indeed reach the bottom. A sounding taken shortly before Stokes's funeral read nearly four miles, at that time the deepest ever measured.

What would happen to the boy's body on its long fall of over 21,000 feet? A two-pound cannon ball would take well over half an hour to reach the bottom. A corpse, far less dense and streamlined, might take hours, assuming it was not attacked and dismembered on the way down. Just as it is impossible at any funeral entirely to suppress anxiety and not wonder, in however fleeting and censored a fashion, exactly what the worms or flames will shortly do, there must have been scientists on deck that day wondering about the effects of pressure on the late William Stokes. (It is most likely that a human body has never been retrieved from such a depth. Although corpses must have been subjected experimentally to enormous pressures to see what happens, the results are presumably buried in the files of naval research institutions.) Wyville Thomson had already written in musing manner: "At 2,000 fathoms a man would bear upon his body a weight equal to twenty locomotive engines, each with a long goods train loaded with pig iron."[11] By now he had got his facts straight about the incompressibility of water.

Any free air suspended in the water, or contained in any compressible tissue of an animal at 2,000 fathoms, would be reduced to a mere fraction of its bulk,

[11]Thomson, *op cit.,* p. 32.

but an organism supported through all its tissues on all sides, within and without, by incompressible fluids at the same pressure, would not necessarily be incommoded by it. We sometimes find when we get up in the morning, by a rise of an inch in the barometer, that nearly half a ton has been quietly piled upon us during the night, but we experience no inconvenience, rather a feeling of exhilaration and buoyancy, since it requires a little less exertion to move our bodies in the denser medium.[12]

At some point the air-containing parts of Stokes's body would have ruptured, principally those of his face, chest and abdomen. The head would not have burst because the cranium contains no air, only incompressible liquids, but the delicate bone honeycombs of his sinuses probably collapsed before water could leak in to equalize the pressure. Sooner or later the chest would have imploded, the broken ends of the ribs coming through the skin. Any air in the gut would probably rupture the abdomen, so if Stokes had been a flatulent boy it would in the end have been his literal undoing. The pressure would also have been likely to cause stress fracturing of certain parts of his skeleton. There might, for example, have been some splitting around the pelvic crest since the abdominal wall is highly compressible whereas the pelvis is not. The same would have applied generally to any structures of finely divided bone (i.e., not solid and thick as in the femur). Stokes would have arrived on the bottom somewhat smaller than he had been on the surface, especially if he was fat, since fat is more compressible than water. The creatures of the seabed would make short work of his flesh, of course, once they had found their way through the holes his rib-ends had poked through the canvas; yet even his skeleton would not last as long as in a conventional earth burial, since bone softens in seawater as its salts are leached out by osmosis. Thus softened, the boy's remains would have crumbled away beneath the pressure.

A vivid demonstration of what deep-sea pressure can do is shown in the experiment beloved by modern oceanographers of sending down with a piece of high-tech equipment an ordinary empty polystyrene coffee cup. It comes back in miniature, a tiny white thimble, all its insulating air cells having collapsed. Yet there seems to be a reluctance to perform this experiment with the body of an animal. I scoured the

[12]Ibid.

Farnella for a ship's rat, hoping that if we could kill a brace we might send them down a couple of thousand fathoms to see what ruptured, but this piece of curiosity was greeted with cries of distaste and accusations of being a ghoul.

In all, the *Challenger* covered 68,930 nautical miles and at the end of 3½ years brought back so many samples of marine plants, animals, seawater, sediment dredgings and corings that it took the next nineteen years to process them. By then Thomson was dead and his place had been taken by his assistant, John Murray. The subsequent report, which by 1895 had reached fifty volumes, has been described as "the most complete expression of man's knowledge of the deep sea."[13] Perhaps as important, the enterprise encouraged similar expeditions by other nations, principally the US, France, Germany, Russia, Italy and the Scandinavian countries. Even Monaco came to hold an honorable position in marine research since Prince Albert I was himself an expert yachtsman and oceanographer who financed his own expeditions. Among his most valuable contributions was a collection of specimens from the intermediate zone which Thomson had thought might be azoic.

Part of the *Challenger*'s achievement was to have laid to rest various misconceptions and to have settled theoretical disputes. Prominent among the latter were post-Darwinian issues concerning living fossils and the Earth's geological evolution. The short answer to the expectation that the deeps concealed living fossils was that they did not. What they revealed was absolute proof that even the greatest depths were neither immobile nor sterile, and that they supported species which, far from remaining unchanged for sixty million years or more, had evolved their own range of special adaptations.[14] In the meantime, other profes-

[13]Susan Schlee, *A History of Oceanography* (1975), Chapter 3. I am much indebted to this excellent work for many details in this and other chapters.

[14]The first time a deep-sea creature was found to fit the fervid category of "living fossil" was in 1938 with the catching of the first coelacanth. There are good reasons for disliking the whole notion of "living fossils," some of which have been noted by Stephen Jay Gould with reference to horseshoe crabs. In taking issue with the meliorist view of evolution and with the tyranny of conventional iconographies—trees and ladders—he objects most to the idea that "old" necessarily means "primitive" or "simple," as if always to imply the superiority of *Homo sapiens sapiens*. In addition he says: "We mistakenly regard horseshoe crabs as "living fossils" because the group has never produced many species, and therefore never developed much evolutionary potential for diversification; consequently, modern

sional judgments were painfully exposed as incorrect. A few years before the *Challenger* expedition set sail, Darwin's friend and champion T. H. Huxley had formulated two hypotheses which became *causes célèbres,* one concerning a notional living fossil of the most primitive kind, the other geology.

In June and July 1857 HM frigate *Cyclops,* while sounding down to 2,400 fathoms, brought up some sediment samples which in due course arrived back in London for examination. Huxley, at the time paleontologist at the London School of Mines, had them preserved in strong alcohol and appeared to forget about them until 1868. On reexamining them after eleven years he found a transparent jelly and became convinced that this was a living slime which carpeted the deep ocean floor, ingesting ooze and forming a rich layer of protoplasm which became a food supply for other life-forms. As such, he gave it the name *Bathybius haeckelii* in honor of the German biologist Ernst Haeckel. Haeckel had been much struck by Darwin's theories and was preoccupied with finding a primitive organism which might provide the missing link between inanimate matter and life. The catchphrase of the day was "abiogenesis" or "spontaneous generation," to describe the belief that living organisms could develop from nonliving matter. (At a microbiological level this was exploded by Pasteur. In the present century an updated form of the idea was floated when amino acids, the "building blocks of life," were generated in the laboratory by imitation lightning discharges in mixtures of gases thought to approximate Earth's primordial atmosphere.) Haeckel was convinced *Bathybius* was the basis of all evolution, the original living matter, or *Ur-Schleim.* Whatever else, it would have to be a subject for investigation aboard *Challenger,* since once Huxley had found and named it, everybody else seemed to be dredging up samples and it was important to establish whether this primordial slime could be found evenly distributed throughout the world's oceans.

For two years *Challenger* found no *Bathybius* and finally the expedition's chemist, John Buchanan, discovered that he could reproduce this

---

species are morphologically similar to early forms" (Stephen Jay Gould, *Wonderful Life* [1990], p. 43*n*). Where the modern coelacanth is concerned it cannot be considered a "living fossil" because no other members of the species *L. chalumnae* have ever been found as fossils. Come to that, "no other species assignable to the genus *Latimeria* has been found as a fossil either" (K. S. Thomson, *Living Fossil* [1991]).

jelly-like substance when he preserved bottom samples in alcohol. He came to the conclusion that this famous protoplasm was no more than calcium sulfate precipitated out of seawater by the alcohol. Thomson at once wrote to Huxley, who promptly, and with immense dignity, admitted the correctness of this chemical explanation and his own error. From that moment *Bathybius* was dead, although several scientists tried in vain to discredit the explanation and its discoverer's retraction.[15]

The second of Huxley's hypotheses also concerned deep-sea ooze, but in a geological rather than biological context. This was his theory of the "continuity of chalk," which, briefly, stated that deep-sea ooze turned into the chalk deposits found on land, so that the continents were formed of the compacted material of the seabed. The essence of this position was the belief of the day that land could only move vertically up and down, for this was long before men like Emile Argand and Alfred Wegener had proposed lateral movement and continental drift. This gave rise to a debate between those scientists who believed the ocean basins and the high continents slowly traded places, and those who thought basins remained basins and continents continents. The *Challenger* soon put paid to the continuity of chalk theory, too. It found that deep-sea oozes were quite distinct chemically from rock formations on land. Besides, geologists had for years been turning up shallow-water fossils in chalk beds on land, showing they could never have been formed in the deep oceans. The upshot was that no evidence was found either for drowned continents or rising ocean basins. This was a particular disappointment to paleontologists, zoologists, botanists and others who thought they needed a sunken land bridge to explain how they were finding close correspondences between species of animals and types of geological formation in otherwise widely separated land masses. They, too, had to wait for Wegener as well as for theories of convergent evolution which explained how unrelated organisms can evolve similar shapes and adaptations in response to similar environments.

[15]An interesting postscript has recently been added to the *Bathybius* story by Dr. A. L. Rice at IOS, suggesting that the seasonal nature of its original collection implies that some of it, at least, could have been detritus of the spring phytoplankton bloom forming a light, flocculent "fluff" on the seabed. This would explain why the *Challenger* failed to find samples, since, being the marine equivalent of thistledown, it simply puffs out of the way of dredges and epibenthic sledges. (See A. L. Rice, "Thomas Henry Huxley and the Strange Case of *Bathybius Haeckelii* . . .", *Archives of Natural History* 2, no. 2 [1983, 169–80]).

~

Behind this gathering of knowledge and the dispelling of misconception and superstition grew a desire to visit the deeps in person. In a way, the development of the submarine from the early twentieth century onward was merely tantalizing, since submarines were incapable of descending deeper than a few hundred feet, scarcely beyond the euphotic zone. Compared with reaching the deeps, this was the equivalent of getting a toe wet. Besides, submarines were war machines, not research vessels, and were far too big. They had the inherent problem of needing to support a large volume of air against great external pressure. This could only be achieved by massive construction, otherwise they would be crushed like a ribcage. Not until the 1930s did a true hero emerge prepared to put his trust in a piece of equipment which Alexander the Great would certainly have disdained.

This was the bathysphere, a name coined by its American inventor, William Beebe. The concept was simplicity itself. A thick-walled steel sphere with a circular entry hatch and a tiny porthole would be lowered like a plummet over the side of a ship on the end of a cable. In effect, it was an eyeball on a string. It could do nothing of itself but carry man's sight into unseen regions. Beebe's accounts of those early dives are in a sense laconic, even though shot through with terrifying images of the physical forces involved. The bathysphere was once sent down empty on a test dive and when hauled up was much heavier than usual. As the first bolts of the hatch were loosened, needle jets of water sprayed out, showing it was partly full and under great pressure. It was clear to everyone that at some point in the loosening process the entire hatch might blow off, yet Beebe and his companion Otis Barton went on unscrewing the nuts with spanners while standing as far to one side as they could. When it finally did blow, the heavy steel plate missed them both by fractions of an inch, flew the length of the deck, humming, and dented a donkey winch. Both men eventually felt the technological problems had been mastered, however, and made a historic series of dives cramped for hours in the tiny space, taking it in turns to squint awkwardly out of the peephole and dictate via telephone what they saw, while a female colleague in the ship far above took it all down in shorthand. A photograph taken on deck of Beebe emerging from the bathysphere shows the physical toll these long dives took. He is barely able to get through the tiny hole, so stiff is he with cold and cramp. There is no clutter of emergency equipment on deck, no officious bustle of rescue teams dressed in special gear; just a thin man with a lined face wearing slacks and canvas shoes being helped out of a steel ball which looks not much bigger than a large mine. In 1934 he and Barton reached the record depth of 3,028 feet off Bermuda.

When Beebe's accounts were not laconic they were filled with the excitement of a man who knows he is seeing things which no other human has ever seen and who responds with the keenest aesthetic pleasure. "If one dives and returns to the surface inarticulate with amazement and with a deep realization of the marvel of what he has seen and where he has been," Beebe wrote, "then he deserves to go again and again. If he is unmoved or disappointed, then there remains for him on earth only a longer or shorter period of waiting for death. . . ."[16] Beebe and Barton did go again and again. After their record descent, Beebe observed, "When once it has been seen, it will remain for ever the most vivid memory in life, solely because of its cosmic chill and isolation, the eternal and absolute darkness and the indescribable beauty of its inhabitants."[17] He was particularly attentive to color and the changes associated with depth. As the bathysphere was lowered through the fathoms Beebe relayed the vanishing of the comforting, warm rays of the spectrum as the colors from red through yellow to green were progressively filtered out, leaving the rest to "chill and night and death." He tried to describe what was left, a paradoxical and strange illumination which was both twilight and brilliant.

It was of an indefinable translucent blue quite unlike anything I have ever seen in the upper world, and it excited our optic nerves in a most confusing manner . . . the blueness of the blue, both outside and inside our sphere, seemed to pass materially through the eye into our very beings.[18]

I quote Beebe because he is both scrupulous and imaginative, and his text is full of small observations which anyone who is thoughtful about the sea will immediately recognize as authentic, such as that you don't

[16]William Beebe, *Half Mile Down* (1934).
[17]*Ibid.*, p. 175.
[18]*Ibid.*, p. 109.

get wet when you dive, only when you surface. He was much taken by the luminous fish that swam past his window and lamented how much more he must be missing. Alexander the Great, scopic prodigy that he was, had watched a fish so huge it had taken three days to pass. Beebe merely records sadly, "A gigantic fish could tear past the window, and if unillumined might never be seen."[19]

Years later in 1949 Otis Barton made a descent off California which increased the world depth record to 4,500 feet, but the day of the bathysphere was done. The next step was taken by Auguste Piccard in his bathyscaphe. This device, which gave him the independence of not having to dangle helplessly on a hawser from a mother ship, was the undersea version of the balloons in which he had been setting world altitude records at the time Beebe was making his first pioneering descents. The bathyscaphe consisted of a pressurized chamber not unlike the bathysphere but slung beneath a large, lightly built tank full of petrol. Since petrol is lighter than water, this was the equivalent of a gas envelope, and because the contents were incompressible there was no need for great strength and weight. Ballast took the bathyscaphe down and, once that had been released at the required depth, the flotation chamber brought it back up again. Piccard emphasized the ballooning pedigree by naming his first bathyscaphe the *FNRS 2*. (The original *FNRS* was a balloon named for the Belgian National Fund for Scientific Research, which had supported the project.) Piccard's account is more technical than Beebe's, mainly because the engineering problems he had to solve were far more complex.[20] Beebe had faced the danger that the bathysphere might have leaked but he was never in any fear of becoming lost entirely. The bathyscaphe, though, might very easily have stuck on the bottom without the remotest hope of rescue, particularly if its flotation tank were holed. Its inventor's story is a triumphant record of doggedly surmounting each new technical problem as it arose. His ingenious answer to the question of how to jettison ballast was a case in point. He needed to be able to dump weight in accurately controlled amounts but the pressure outside the hull precluded any mechanical device passing through it. Apart from the additional danger of leaks, boring holes in the steel shell (which was cast and milled in two hemi-

[19] *Ibid.*, p. 221.
[20] Auguste Piccard, *Au fond des mers en bathyscaphe* (1954).

spheres) would weaken it. His solution was to use steel instead of lead shot as ballast and to jettison it through a hole around whose circumference were electromagnets. By pressing a switch inside the cabin Piccard could energize the magnets outside and lock the balls solid, blocking further release.

The bravery of these men seems extraordinary now, and it would be churlish to complain that Piccard was no Beebe when it came to describing what he saw when he went down. The fact is, his brilliantly engineered invention took him down very much farther. In 1960 the *Trieste,* the latest version of the original bathyscaphe, reached the bottom of the Marianas Trench at 10,916 meters, some 35,800 feet or better than 6.75 miles. This is as deep below the ocean's surface as the highest-flying passenger aircraft leaving its white contrails is above it. Effectively, it was the deepest point in the oceans. Quite possibly there are places where this is exceeded by a few tens of meters, but to all intents and purposes man had gone as deep into the oceans as he ever would, just a century after Darwin's *The Origin of Species* was first published. Since then, the technology of deep-sea descent has become ever more refined and flexible, permitting proper, if limited, exploration. As with air travel, systems have become very much safer. This is by no means to belittle the courage of men like Robert Ballard, since the possibilities for disaster are still endless, miles beneath the last glimmers of daylight and with prodigious pressures ready to slam shut the tiny bubble of living space at the first sign of a weakening rivet.

~

The pleasure Beebe took in the luminous fish he saw was mixed with wonder at this evidence of a rich and entirely alien way of life where ambient darkness was as little problem as either cold or pressure. More than sixty years earlier when aboard *Porcupine* shortly before the *Challenger* expedition, Wyville Thomson had noted the light given off by coelenterate fauna such as gorgonians and sea pens brought to the surface in the trawl, marveling that it should be bright enough for him to be able to read his watch by. On one occasion what he saw gave him a glimpse into that illuminated underworld.

The trawl seemed to have gone over a regular field of a delicate, simple Gorgonid. . . . The stems, which were from 18" to 2 ft. in length, were coiled in great hanks around the beam-trawl and engaged in masses in the net; and as they showed a most vivid phosphorescence of a pale lilac color, their immense number suggested a wonderful state of things beneath—animated cornfields waving gently in a slow tidal current and glowing with a soft diffused light, scintillating and sparkling on the slightest touch, and now and again breaking into long avenues of vivid light indicating the paths of fishes or other wandering denizens of their enchanted region.[21]

Much research has gone into the bioluminescence of different marine organisms, a subject made more complicated because the light has no unitary function. It seems that flashes of luminescence may be used variously as a defense, to entice prey, and as a sexual display. It may be seasonal or constant. It is even thought that some creatures may adroitly vary the wavelengths they emit, thereby using light itself as a method of camouflage. This is the same principle as the red colorations used by deeper reef creatures to make themselves look gray and stone-like, though beyond a limited depth there is little point in talking in terms of color. The theory goes that an animal could camouflage itself by emitting low levels of light if it exactly replaced that lost by absorption on its upper surface. The light would have to be of precisely the right strength, at the right wavelengths and of the right angular distribution (since below about 400 meters the remaining light falls vertically and is no longer refracted at other angles). There is increasing evidence to support this theory, and certainly the eyes of many creatures of the deep-twilight and lower zones are highly sensitive to light and to the subtlest variations in its intensity and wavelength. Animals such as squid and hatchet fish use amounts of daylight which would appear indistinguishably black to human eyes in order to regulate their vertical migrations.

The discovery that vast numbers of animals rise to the upper waters at night and return to the depths during the day was surrounded by secrecy in World War II. Three scientists experimenting with sonar aboard the USS *Jasper* in 1942 had found a layer in the water at between 1,000 and 1,500 feet from which echoes bounced as if it were solid. This was not made public until 1946 because it was thought an enemy submarine might take advantage of the layer by hiding beneath it. In 1945 the Scripps Institution of Oceanography found that this layer

[21]Thomson, *op. cit.*

moved up at night and down during the day and concluded it must be alive. Now known as the Deep Scattering Layer, its movement varies seasonally and from place to place. It consists of huge numbers of small animals migrating punctually up and down the water column, some by as much as 1,500 or 2,000 feet. The DSL is probably the chief cause of bogus sonar contacts, and many a ship has reported "lost" land lurking just beneath the waves where later investigators have found only thousands of feet of water.

The deep triumphantly disclosed the consequences of Darwin's ideas of natural selection in that the often bizarre colors and shapes of abyssal fauna emerged as exquisite adaptations to extreme circumstances. The "azoic" theory had betrayed as nothing else the limits of understanding of the nature of life, and how erroneous all judgments were when based solely on human considerations of what might constitute a livable environment.

Where "monsters" are concerned, they may yet be found, although it is unlikely, owing to scarcity of food, that they will be from the very deepest parts of the ocean. However, if the Kraken is mythical, the giant squid is not. Huge specimens and fragments of even huger have occasionally surfaced. It is clear from measuring the sucker marks on dead whales that immense battles must take place in the middle deeps involving squid of a size never yet seen. The great mass of the oceans remains unexplored, even as the contours of their beds are electronically surveyed. Their waters must hide many species strange to taxonomy, but this is hardly surprising. Letting down nets here and there may catch few creatures with acute sensory equipment and evasive powers. The world a mile or more down keeps its secrets well, with neither victors nor victims necessarily leaving the least trace of their lives. As regards the deepest trench faunas, there has been relatively little recent research because most of the effort and money has been directed toward studying the vent communities around "black smokers," which have the required glamour to attract funding.

As to geology, the seabed turns out to be of great use in climate modeling. It is possible to weigh the atoms of oxygen trapped in fossil shells brought up in sediment cores and determine what the temperature was when the creatures were alive. Such cores have also yielded information about monsoons and glaciation. It seems the present pattern of monsoons only started some ten million years ago, and a theory has been put forward that they have been directly influenced by the vertical uplift of land masses (as a result, one should point out, of horizontal movement elsewhere). In the last million years alone the Himalayas have risen over two kilometers and it now seems likely that winds and precipitation have been directly influenced by this uplift, much as the construction of a groin or breakwater can lead to the silting up or scouring of an adjoining bay.

"There rolls the deep where grew the tree. . . ." The Ice Age locked up enormous volumes of water during the Pleistocene, when what today is known as Dogger Bank in the North Sea emerged as land. It was boggy and forested and became full of men hunting animals with flint weapons, chasing deer and bear and wild ox among the willows and birches. None of this was known until the nineteenth century, when widespread trawling started and to their surprise fishermen discovered a lumpy plateau almost the size of Holland lying only sixty feet below the sea's surface. They inferred that this had once been land when they began netting bones and axeheads and moorlog (a kind of peat). The waning of the Ice Age, that era's equivalent of the greenhouse effect, brought an endless close season to the Pleistocene hunt. There must have been a long, mournful period of many centuries as the ice melted and the sea level began to rise again to turn this land between East Anglia and the Netherlands into an archipelago, dozens of scattered islands with heterogeneous collections of hyenas, woolly rhinoceros and mammoth struggling for survival on ever-decreasing patches of territory. Then, at length, nothing but the deep. A mere 50,000 years ago and the forests of Dogger would have been visible from what is now the coast of Lincolnshire. Tennyson, fast in the grip of transience and loss and Charles Lyell's bleak discoveries, had no need of them to complete his vision.

> The hills are shadows, and they flow
>   From form to form, and nothing stands;
>   They melt like mist, the solid lands,
> Like clouds they shape themselves and go.[22]

---

[22]Tennyson, *In Memoriam*, cxxiii.

2. THE MONSTERS WITHIN

On the coast of a Philippine province there is a small town. On the landward side of the road, set well back among coconut palms and jasmine, is a white-washed church with a green tin roof, only one of several civic buildings including an elementary school and an abandoned health center. A legend surrounds this church, one known to every fisherman in town and to every boy who ever jumped off the little coral pier clutching a speargun. The legend *underlies* the church rather than surrounds it, for the story goes that there is a passage leading from the sea to a cave deep beneath it which is the lair of a giant octopus. There certainly is a fissure in the thick cap of fossil coral which covers much of the volcanic basalt of the island's coasts. Its mouth lies about twenty-five feet below the surface at high tide and at night a powerful underwater flashlight shone nervously in reveals no end to its interior.

In the absence of scuba gear there is nothing to be

done, since only a madman attempts to explore a submarine cave with a chestful of air. There would be nothing to be done in any case. Whether or not one has been worked on by the legend, this particular depthless black slit does exude a peculiar aura of menace. The water around the mouth is always several degrees colder than elsewhere and very few fish appear ever to venture in. If there really were some great monster lying tucked away inside, a good deal of food would need to swim unwarily in for it to survive. On the other hand it may be that one way or another much of the town's drains seep into this crack and the creature survives on ordure. It might even have grown so fat it could no longer leave if it wanted to and is bottled up in its coral crypt. The thought of swimming up a sewer to confront a trapped monster is another good reason for not making the attempt.

These are all excuses, of course. The fact is, I am afraid of the place and so is everybody else. From the sunlit surface above its opening one can look across the road through the palms and see the church's corrugated roof. If the fissure really does extend that far it must be at least 100 meters long. Up in the brilliant daylight the whole legend looks different. The idea of a demon Kraken lying in its lair beneath a church is too naively Filipino, too redolent of Christian mythmaking to be more than the embroidery with which the credulous have ornamented a freak of local geology. Nevertheless, I am not going inside.

Only a mile or two down the coast and not far out to sea the reefs drop suddenly into ultramarine depths. By swimming out over the shallow corals it is possible to pretend one is low flying, hedgehopping above rough coral terrain, an illusion strengthened by soaring out over this great abyss. So abrupt and powerful is the effect one may even feel one's stomach drop. It was here, over the years, that I would practice seeing how far down I could swim, to set my own private record. Soon I knew every ledge on this cliff face, each downward step of the agonizing but exhilarating journey. I knew each level by its peculiar feature—a coral outcrop or eccentric sponge—and also by its ambient light. I knew as well as Beebe where red became gray, where my blood looked dark green in the water. This series of terraces now seems a lunatic vertical *via crucis*, every step gained representing pain, but it also stands as a chill measure of aging. I can no longer reach the very deepest of the shelves I once touched, raising a confirmatory plume of silt still visible from the surface like a triumphant smoke signal impressively far below. I might again, I tell myself, but only if I lost weight, since fat is buoyant and means one has to burn up more oxygen to drag it all downward. Since I can no longer measure my own record, I have to estimate it as between eighty-five and ninety-five feet, rather less than the average local teenager can manage when harvesting big white sea cucumbers.

It was here, perhaps to spite myself, I tried "riding the rock" instead, or using a weight to pull me down. This is how world free-diving records are set. Nobody labors to swim down; they ride the rock and somewhere beyond 300 feet let go and hope their still uncollapsed lungs contain enough air for them to claw their way back to the surface. Taken to such depths it is a dangerous sport, but I had no intention of going that far. In the event, the whole business felt faintly embarrassing. There is something foolish about loading rocks into a dinghy, rowing out, and jumping overboard clutching them to one's chest. The first was too heavy and took me down so fast I could not "clear" in time and the pain in my ears made me let go at about thirty feet. The next took me down rather languidly, and it was a pleasure to see ledges I had fought to reach drifting upward past me like floors in a descending lift. With a subsequent rock I passed my own record and was pulled onward into unknown territory. I do not believe I ever went farther than about 160 feet. There was something disagreeably inexorable about the downward tug. It was not as if one doubted for a moment that one's arms would release the weight before it was too late, but it had something to do with increasing pressure and deepening gloom which I had never experienced without breathing apparatus. Perhaps because the motion was entirely vertical and swift, one imagined dissolving like a meteor, leaving a trail of silver bubbles, soon to be worn away to nothing by the rasping caress of the sea.

Beebe had written: "The only other place comparable to these marvelous nether regions must surely be naked space itself . . . where sunlight has no grip upon the dust and rubbish of planetary air." The exploration of space and of the deep sea have obvious things in common. Both require venturers to be supplied with complex life-support systems and defended against extreme ambient pressure, whether positive or negative. At a mythic level, however, there are important differences, many of which—in the case of deeps—have to do with the dark.

The famous and fatuous opposition of light and darkness is pre-Socratic in origin, only one pair of many made up of a "noble" element (right, above, hot, male, dry, etc.) and an "ignoble" (left, below, cold, female, wet). By the sixth and fifth centuries B.C. the faculty of vision and the attributes of knowledge had run together in the Greek word *theorein,* meaning both "to see" and "to know." Knowledge was henceforth a register of vision. Ignorance therefore becomes a lack of knowledge predicated on objects not being visible, so darkness equals ignorance. In turn, the dark becomes a source of fear as if a knowledge of visible objects were the only defense against terror and anxiety. By the eighteenth century the light of reason stood for the banishing of primordial fear: literally, enlightenment. Superstition as a concept is a product of eighteenth-century topology.[23] Where the ocean's deeps are concerned several other dualities operate as well, such as up/down, lightness/pressure, outward/inward and future/past. To go into space is in some sense to go forward; to go down into the depths is at a psychic level to regress.

Why should this be? Space travel is "going forward" in the obvious sense that it involves technological "progress," but so does deep-sea exploration. It is as though *Homo* viewed himself in spatial rather than temporal terms, as if his history had been one not of eras and dynasties so much as of steady territorial expansion. Maybe the whole of human history might be rewritten, leaving out dates and measuring instead the boundaries pushed outward by tribes on their way to becoming nations, by earthlings as they stake out their claims to colonize the solar system. Yet even with nations claiming EEZs and seabed rights it never feels an appropriate choice of cliché when journalists call the ocean depths "the last frontier." As always, the sea is really less connected with space than with *time,* as if there were a correlation between going deep and going back. Thus the deeper one went, the more primitive would be the life-forms encountered, the more prehistoric and inchoate.

This must be a comparatively recent idea, post-Darwinian, at any rate, and taking into account a popular version of Victorian scientists' excitement on learning that the deeps were not azoic. The finding of the first coelacanth would have strengthened this, as does every fresh "sighting" of the Loch Ness monster. Legends of monsters and sea serpents are at least as ancient as the written word, but presumably it is only after the

[23]This passage originates in a lecture given by Mark Cousins at the Architectural Association in London on 23 November 1990.

mid-nineteenth century that they begin to be depicted as prehistoric and corresponding loosely to fossil forms. The Loch Ness monster is almost invariably spoken of nowadays not as some unknown species of sea snake or eel but as a saurian of prehistoric type. Since this is what people wish to be there, it is faithfully confirmed by all the "sightings." It is thus a true remnant of a misapprehension by nineteenth-century science.

Myths of space travel do include visits to worlds at an earlier stage of evolution than our own. Yet even these are often in "obscure" back-waters of space as if in the scriptwriters' imaginations space did corre-spond to a vast ocean in which the most developed regions tended to be those appearing from Earth most brightly lit. ("Rigel Concourse" in Jack Vance's stories is a good example, Rigel being a pure white, first-magnitude star. This is exactly where one would expect to find our outwardly bound pioneering descendants rather than huddled around some dismal cepheid variable out in the galactic sticks.) All this apart, the creatures most commonly associated with space operas as well as with UFOlogy are of an intelligence superior to ours: and with the waning of American paranoia about communism they tend to be less and less bent on kidnapping and brainwashing. Nowadays space aliens may well incline toward the god-like, beings from whom we might acquire knowl-edge, enlightenment, light itself, before it is too late.

The mythology of our own Planet's oceans is the polar reverse of all this, so much so that the nether world sometimes seems hardly part of the Earth at all. It is worth examining this from the popular standpoint for a moment because it shows how the concept of "the deeps" relies on a jumble of associative ideas. Far from being likely to find enlighten-ment the farther down we go, then, we expect to meet ever-dumber creatures. Moreover—exactly opposite to actuality—we envisage them near the bottom as still bigger, more terrifying in their mindless strength, and *uglier.* . . . In fact, monsters. To this extent they are remarkably similar to the nightmare creatures of the unconscious: tentacular horrors which enwrap and bear their victims down and down to lairs where, in due time, begins the business of the hideous rending beak and saucer-like eyes. The very gradations of sleep itself seem to suggest a vertical descent into annihilating depths, the deepest levels of sleep being those of oblivion. The levels of dreaming, like the layers of the ocean which can support the biggest life-forms, lie nearer the surface. In any case, by descending into the sea we would expect to meet the monstrous rather than the divine. Gods are the last things we would imagine finding in the deeps. It is no accident that even the men we encounter tend to be people like Captain Nemo, ominous whichever way we read his name. Astronauts have claimed close encounters with a Supreme Being, but never deep-sea divers. Nor should we be surprised. Superior beings are by definition on top, while only the inferior can lurk below. The deeps also remind us of where we suppose we originally came from, what we have left behind. Going back thus to our genetic roots rather than to the sunlit idyll of Eden is a disquieting affair. Did we not abandon our ancestral dark by crawling toward the light?

No; we did not. The sea, to its dwellers, is not a dark place. With exceedingly acute eyes perceiving low levels of light and complex codes of bioluminescence; with sensitivity to sounds, smells and minute pressure differentials far beyond the spectrum of our own senses, it is as pointless to speak in crude human terms of "light" and "dark" as it would be when speculating about what a bat sees. A bat "sees" with its ears with great precision and at speed. In short, there *is* no such thing as darkness. It exists only in the perception of the beholder. Vision does not depend on light.

To these "oppositions" and their associations (up/down, above/ below, superior/inferior, heaven/hell) should be added striving/sinking, where the first generally implies upward aspiration and self-betterment and the second is redolent of slummocking on a downward path, of Jack finding his own level while still undrowned. "Sinking" is also used to describe wretched people glistening with sickness on their deathbeds, as if their problem were only one of weakness and they could no longer resist the force of gravity tugging them down toward their graves. That there might be something subtler at work than these pairs of opposites is suggested by the Latin word *altus,* which can mean both high and deep (as it does still in Italian and where *l'alto* means "the Deep" in an oceanic sense which also lingers in the English phrase "the high seas"). In the Freudian unconscious, at least, such an idea would not embody a contradiction because there are no contradictions in the unconscious. Entirely antithetical and mutually canceling propositions can exist simultaneously with not the slightest difficulty.

Perhaps, then, the least strange thing about the Deep is the degree to which it has retained its psychic force, its sonorous and chilling stateliness, its amalgamation of height and depth, of gulfs of space and of time.

Almost no matter what is done to the oceans, however much they are explored and exploited, even ravaged and polluted, the Deep surprises us by its resistance to contamination. In this respect it resembles the Moon, which still feels much the same even though we know its dust bears the frivolous prints of cleated boots playing golf. The fact is, it was a different moon on which the astronauts landed, just as it is a different deep which GLORIA deafens with its sonar signals and whose silt is scarred by remotely controlled sleds gathering the sort of things a sled would gather. Neither Beebe nor Piccard nor Ballard ever visited the Deep. They reached various depths, even the ocean bed, but they carried the Deep within them. It is not a space to which there is physical access. Yet an air of mystery, no matter how slight, still surrounds objects retrieved from the depths, even beer bottles and polystyrene cups lowered by the curious. People like to touch things brought up, such as hoppers full of nodules. They like to feel the chill of eons before it fades, just as they like to handle meteorites and moon rocks. If the ocean vanished tomorrow its mystery would not be found in the sum of its creatures flopping and dying and rotting on its bed. It exists elsewhere altogether, as Tennyson well knew when he capitalized on its high melancholy to express his grief over Arthur Hallam's death, hidden and heightened in a transition: "From the great deep to the great deep he goes."[24]

[24]Tennyson, *Idylls of the King,* "The Coming of Arthur," 1.140.

3. DATING THE EARTH

Willliam Buckland's *Bridgewater Treatise* maintained that the presence of fossils embedded in sedimentary rocks was definitive proof of Noah's Flood. This work, really a series of lectures, was published in the 1830s, the same decade as Charles Lyell's *Principles of Geology*, whose radical conclusions could scarcely have been more different. By Buckland's day the question of the Earth's age was keenly debated. The natural sciences had evolved to the point where a more serious answer was needed than Mosaic chronology could provide. The supremacy of the biblical version of Earth's creation had already been challenged a century earlier by Newtonians like the comte de Buffon (1707–88). Eighteenth-century science had gained enough insight into geological processes to enable it not so much to speculate about how old the Earth was but to wonder how all the vast, slow procedures of erosion and sedimentation and the laying down of fossil beds could have been squeezed into the mere 6,000 or so years apparently allowed by Scripture.

It is possible to argue that until Christianity there was no such thing as time. That is, until after the life of Christ people's notion of time would have been largely cyclical, based on the regular recurrence of seasonal and astronomical phenomena. Longer periods were simultaneously precise and vague, being measured by carefully preserved familial dynasties such as the lengthy genealogies in the Old Testament. To small rural societies the earth was unchanging, as old as legend, as old as their creation myths. It was not possible to apply any external timescale to it because one did not exist. This view of time changed radically with the coming of Christianity. Suddenly, time stopped being a slow, circular continuity and became an "arrow," linear, flowing in a single direction from Creation to the Last Judgment. Christ had come as a man in fulfillment of a prophecy, and as a man he had died. He could only do so once, so his life was just an episode—though a momentous one—in a single temporal trajectory which was carrying all mankind with it toward the great Millennium.

Once this idea had taken root, theologians began examining the Judaic scriptures with new motives. They were now less interested in the genealogies as evidence of pedigree than as ways of calculating how many years had elapsed since God had created the world so lately visited by his son. Theophilus of Antioch (115–83) put Creation at 5529 B.C.; Julius Africanus (200–250) at 5500 B.C. This was eventually reduced by Martin Luther (1483–1546) to 4000 B.C. and finally Archbishop Ussher (1581–1656) produced a date of scholarly precision, 4004 B.C., which stood as the problem's definitive solution. This date gained such wide acceptance it was printed in most English bibles with chronologies. It can be found even today in some fundamentalist bibles.

This firm date of 4004 B.C. had all sorts of consequences for the way in which people thought and looked at the world and soon produced intolerable strains in the burgeoning natural sciences. Even archaeologists began worrying about how societies as complex as that of ancient Egypt could have evolved so quickly. To some it was clear that the oldest pyramids dated from around 3000 B.C., only a scant thousand years after the beginning of the World According to Archbishop Ussher. Geology, meanwhile, had progressed to where the publication of Lyell's *Principles of Geology* in 1833 made it clear it was no longer necessary to invoke six-day acts of creation or great flood catastrophism in order to explain the Earth's structures. All that were needed were the ordinary processes which anyone could observe, plus almost unlimited quantities of time. Once it was assumed that fossil beds could be millions of years old instead of a maximum of 6,000, it created a sort of conceptual breathing space in which many things suddenly began to make sense.

No matter how horrible or absurd their positions may be, there is always an element of poignancy about diehards. It was therefore both comic and pitiful to see a scientist like Philip Gosse confront evidence such as Lyell's and be unable to relinquish his own fundamentalist position. Instead, he went into contortions which effectively destroyed him, bringing down on his fervent and well-meaning head the ridicule even of churchmen. In 1857, only six years after his school zoology textbook, he published *Omphalos,* which, as the title implies, had as its center of gravity the question of the navel. Adam must have had a navel because God had created him as the genotype of the human race. Since Adam had no mother, his navel was surely more exemplary than functional. Similarly, had Adam cut down one of the trees in the Garden of Eden he would have found annual growth rings, although they had been there only days. To deal with this problem Gosse claimed that some things were "prochronic" or "pre"-Time. Adam's navel and the tree rings were prochronic. Because the first chicken had been created fully fledged, the putative egg from which it hadn't in fact hatched was equally prochronic. So also were fossils. God, for reasons best known to himself, must deliberately have "salted" the Earth with fossils in order to make it seem older, perhaps, and maybe even to test the faith of later scientists, especially geologists.

The derision which greeted this pretty notion completely baffled poor Gosse. The Reverend Charles Kingsley, whose *The Water Babies* showed he was himself quite capable of imaginative extravagance, was particularly forthright. In his own *Glaucus* he commented bleakly: "If Scripture can only be vindicated by such an outrage to common sense and fact, then I will give up my Scripture, and stand by common sense." Gosse was not completely on his own, of course. William Buckland went on thinking that the Flood was a perfectly satisfactory explanation for fossils. But Gosse was a well-known figure, and his flounderings in the name of science made him something of a lightning conductor. Others, too, were disconcerted and depressed by a rationalism which seemed unaesthetic as much as irreligious. Ruskin wrote plaintively in a letter of 1851: "If only the Geologists would leave me alone, I could do very well, but those dreadful Hammers! I hear the clink of them at the end of every cadence of the Bible verses."

In the mid-nineteenth century what science badly needed was a reliable temporal yardstick by which to obtain a sensible age for the Earth, and in the following 100 years many were proposed and tried. Having gone down a coal mine and noticed the higher temperature, Lord Kelvin concluded that it ought to be possible to work backward using the temperature gradient and extrapolate a date for when the Earth was in its original completely molten state. He suggested an age of twenty to forty million years, later putting it at 100 million years. Some geologists considered sedimentation rates, using the annual deposits known as varved layers rather like tree rings. William Beebe thought salt would prove a reliable measure. In *Half Mile Down* he wrote: "The chemical composition of blood, both in the constituent salts and their proportion in solution, is strangely similar to that of sea water." The only difference was that our blood is three times less saline than the sea. "So all we have to do is calculate back and find the time when the ocean was only one-third as salt as the present . . . and we will know [the exact moment of] our marine emancipation." True, this would not be the age of the Earth itself, but it was still a useful date to fix. Unfortunately it now seems that the ocean's salinity has never varied very much and that prehistoric seawater was not greatly different from modern. This also put paid to the idea for calculating the age of the oceans by assuming they started as fresh water and working out how long it would have taken to leach out that amount of salt from the Earth's crust.

It was only with the discovery that radioactive elements decay at strict rates that a reliable geological or cosmic "clock" was found. Nowadays the age of the earth is usually given as 4.7 billion years, but even this is not final. In fact, all cosmological dates are under constant revision and may always be so. It is unclear whether the construct of time's linear "arrow" can survive quantum imponderables, but in short timescales at least it seems to point to a cheerless future for the human race. The discovery that the universe is not static meant that the Earth cannot last indefinitely, though the precise manner of its end is conjectural. In the 1930s the theory propounded by Sir James Jeans and Sir Arthur Eddington of the "Heat Death" of the universe was received with the kind of glum consternation that represents less an absolute belief (for few people understood the physics) than a general shift in the recognition of what is plausible. Public moments like these can be looked back on as forming yet another step in a sequence which included the great debates centering on the writings of men such as Gosse, Darwin and Lyell. The trend was inexorable. The human race was slipping out of the hands of God and into a quite other universe determined by the second law of thermodynamics.

Nowadays *Homo*'s fate is commonly thought to be even worse than that, being in his own hands. Death by nuclear destruction, environmental pollution, global warming or in a demographic gridlock of overpopulation are fluently forecast by prophets of one sort or another. Serious minds can be found expending as much effort in predicting man's future span as in trying to date his past. In the face of all this the choice of a dignified intellectual stance seems limited. The fossil record might be fragmentary, but its message is all too plain. Mass extinction for one reason or another has occurred many times, no doubt more often than we know. It could happen again tomorrow without violating a single natural law. Since the mindless optimism of religion cannot qualify as a serious position, rational man is left in possession of his true intellectual birthright, an exalted stoicism. Bertrand Russell, having read Jeans and Eddington's theory, put it succinctly:

"Only on the firm foundation of unyielding despair can the soul's habitation henceforth safely be built."[25]

---

[25]Quoted in John D. Barrow, *The World Within the World* (1988).

"From henceforth thou shalt catch men." Quota-free fishing on the Sea of Galilee.

The sun is definitely lower now. The freezing thermocline in whose upwelling the swimmer was briefly caught has moved on. He is thirsty and weary and finds himself becalmed in an edgy resignation. He has done his purposeful but unrewarded swimming about. Now he wants to preserve his strength for staying alive as long as possible. He is already thinking ahead to another day's floating, taking it for granted that he will survive the night.

For the first time he is considering the possibility of rescue. He has abandoned the idea of finding his boat. He accepts that his directionless first attempts to search for it are more likely to have separated him still further. Even if he did happen to be looking in the right quarter when stern or prow or the tip of an outrigger reared up on a wave, there are surely too many intervening waves for anything to be visible now.

He does have a plan of sorts, if that is not too intentional a word for such an impotent state as his. At nightfall, he knows, this area becomes a major local fishing ground. True, many of the boats will have engines over whose unsilenced blatter his shouts may not be heard. But many of the poorer fishermen stop their engines to save fuel and just drift, while the poorest of all will come out here under sail. Since sound travels well over water the swimmer has high hopes that someone will hear.

In the meantime he is once again examining the sunlit depths on the extreme off-chance of rescue from another source. He has heard legends of dolphins helping shipwrecked mariners, of a strange bond which sometimes leads them to aid distressed humans, even occasionally towing them to safety. The sea is empty, however. It seems to him it is a long time since he has even glimpsed a dolphin, several weeks at least. He can remember when it was hardly possible to look at the sea for five minutes in these parts without their breaking the surface, leaping in pairs. Only three or four years ago he would probably have been surrounded by the curious and playful creatures. Now there is nothing. The sea is empty even of their squeaks. The swimmer knows their absence is most likely due to the very fishermen at whose hands he is hoping for deliverance. Why should any remaining dolphin come within a mile of him? Of what use now to invoke "strange bonds" in so self-interested a fashion when the deal had always been so cruelly one-sided?

VI

FISHING
AND LOSS

1. FISHING AND LOSS

Modern commercial fishing is a strange, hybrid profession. It affects to be part of the hunting tradition, while thinking like a form of agribusiness. It really resembles neither. There is no element of true hunting left in it, since the prey is detected by electronic fish-finders. At most there is a certain amount of searching to be done in areas where experience suggests it may be worthwhile looking. Nor is it like agriculture, since no farmer ever reaped a harvest without sowing so much as a grain. Only gatherers do that.

There is nothing suggestive of nomadism about the crew of the *Garefowl* which musters at the wharf in Fraserburgh harbor, Scotland, early one morning. It is 2:00 A.M. in April with a northeasterly wind shivering the lights in the black harbor water. There are six of us. Two have arrived in cars, while the rest, like the skipper and myself, have walked out of the warmth of solid stone houses trailing plumes of breath. The town is still.

Yet there is a certain alertness about the closed doors and windows, as if they were quite used to opening and shutting behind the menfolk at all hours of day and night, ready to show anxious faces or disgorge rescuers at the firing of a rocket. Donald and I have walked past a huge stack of empty fish boxes, blue plastic crates designed to hold ten or more stones (140 pounds) of fish. I thought of Honolulu harbor with its coffins, but the chill black air blowing down from beyond Norway belongs to a planet which does not contain anything as frivolous as tropics or Hawaiian languors. Fraserburgh is a typical fishing community and the *Garefowl* a typical small trawler, high-prowed, blunt, a mere fifty-six feet long. There is a lot of wood in her. She is quite unlike the high-tech monsters moored elsewhere in the port, as much factories as ships and bristling with arrays of aerials, antennae, DF loops and radomes.

Donald takes us out quite fast through a maze of high stone jetties and harbor walls. Because of the upward-tilting bows there is very little forward vision. Gray, weed-streaked blocks of stone slide past a few feet away, then a white lighthouse, and we are heading for a place where we "might pick up a few haddock." This is a good three hours' sailing, but as we may be out for three or more days there is no great hurry. In the intervening time Donald tells me his woes. He sits on a chair with his feet propped on the boss of the wheel. From time to time he glances up at the illuminated compass set flat in the wheelhouse ceiling and twiddles the spokes with one foot. To his left glow two screens: the radar and the brightly colored display of a Koden fish-finder. Directly before him are a Decca Navigator and Track Plotter.

"Whatever we find," he says, "you'll have a ringside seat at a calamity. If we catch a lot, ten to one we'll be discarding most of it as undersized. If we catch nothing it'll only go to show. They're a desert, these waters. A desert. You'll see."

Conventional wisdom expresses the problem as too many boats chasing too few fish. Indeed, reducing the number of fishing boats in the North Sea, even if it means governments having to buy them out ("decommissioning"), is supposed to be one of the main objectives in Brussels. Immensely complex bureaucratic yardsticks are variously applied: tonnages of boat, horsepowers of engine (expressed as kilowatts, however), the numbers of licenses issued for boat owners to catch "pressure stocks" (the most threatened species). These aside, the mass of regulations governing every other aspect of fishing is a bureaucrat's dream. Those prescribing net mesh sizes draw bleak snorts from Donald, while a new mandatory eight-day tie-up period causes an outburst.

"What the hell is a small trawlerman like myself supposed to do? Last month I was at sea five days. How can I make a living like that? I won't work weekends, why should I? My father told me on his deathbed that if ever I worked on Sunday I would have nothing but grief and die poor, and I believe him. So that's eight days gone out of the month. Then these new EEC regulations mean I have to go down to the Fisheries Officer and give him twelve hours' notice of when I want to take my tie-up. That means your boat is tied up in the harbor for eight consecutive days, excluding weekends. How can any man reckon in advance when to take his eight days? It depends on the weather, doesn't it? Or repairs. So as it happened the weather was awesome as soon as my tie-up ended. Out of the sixteen days I *might* have fished last month, I could only get out on five. Probably we'd all of us accept eight days off a month, even ten, provided we could pick and choose and didn't have to take them all in a block. It's killing the industry and driving us to crime."

I imagine bootleg whisky or drug running and he is angry that I am taking what he says too lightly. He thrusts a recent copy of *Fishing News* at me, then snatches it back at once and begins to read from it by the light of the various screens.

"And this is by the editor himself, mind," he says. " 'It's impossible for any skipper with a boat over eighty feet to stay viable now without breaking at least one of the three main rules—misreporting, using a smaller mesh or making illegal landings. Impossible on the quotas we have now. We're being forced to be criminals.'[1] Crazy, you see. Any night you can go down the port in Fraserburgh and Peterhead and watch boats illegally landing catches way over the quota straight into lorries. You can wager there's never an FO there. I don't know where they all go to. They just vanish at the right moment."

Having reached the grounds, Donald loses way to shoot the net. By the light of the stern spotlamps it flows overboard, followed by a series of plastic floats the size of skulls. The two trawl doors are also swung into the sea, heavy iron rectangles whose drag in the water will keep the mouth of the trawl open. Soon it is all at sixty fathoms, the speed

[1]Tim Oliver, *Fishing News*, 26 April 1991.

increases to four knots and there is nothing to be seen but the twin tow cables thrumming the length of the boat from the winch for'ard and out into the darkness astern.

There will be little to do for the next four hours. Ordinarily the men would crawl into the wooden hutches down below to sleep or read, but nobody is tired yet. Dawn is only an hour or two away. Tea is brewed and the wheelhouse fills with men who have sailed with each other for years, have known each other since childhood. Two of them grew up in the same street as the murderer and their contemporary, Dennis Nilsen.[2] It is like sailing in the company of great auks, an extinct species. They are rationally angry at the botched and muddled decisions by successive governments and the EC; underneath is a more passive note of sadness that their livelihood is coming to an end and with it the long tradition of an ancient community.

"There's still a mess of money to be made out here," points out Graham. "You've seen all the new housing outside Fraserburgh? All owned by fishermen, if you can call them that. Some are little better than boat drivers. They know nothing about the sea but ten years ago they liked the look of all those grants London and Brussels were handing out like blank cheques, so they ordered up boats costing millions and then fish prices doubled twice over and they got rich. Mind you, they're still paying off for those boats and all that flashy tackle. But they think in far bigger terms than men like us can. With that gear they're landing £50,000 [about $90,000] worth of fish at the end of a week's trip, having broken every law in the book. They've rigged the nets so the diamond mesh is squeezed practically shut. They might have put blinders on into the bargain—that's another net covering the first. Or if they'd had a governor on their engines to de-rate the horsepower they'll have broken the seal half an hour outside Fraserburgh. No problem. They know how to put a lead seal back on so the Fisheries inspectors

[2]Nilsen, a distant relative of Virginia Woolf, frequently betrays in his prose and verse a poetic sensibility. This is clearest when he describes a lonely childhood beside this harsh northern sea. His personality, like Tennyson's, was marked by its proximity. In his prison cell, shut off from the sound of gulls and waves, he writes: "I am always drowning in the sea . . . down among the dead men, deep down. There is peace in the sea back down to our origins . . . when the last man has taken his last breath the sea will still be remaining. It washes everything clean. It holds within it forever the boy suspended in its body and the streaming hair and the open eyes" (quoted in Brian Masters, *Killing for Company* [1985]).

can never tell. Then they'll probably land their catch illegally. It's a joke, ken? And if they're caught, what's a £5,000 maximum fine to them? It would put us out of business, but not them. Some of those families run three or four boats."

"We're not saints," Donald puts in. "Don't think we never bend the rules ourselves now and then. We have to, otherwise we'd starve. There's nothing left out here for us small folk. It's all been swept clean. Winnie Ewing was right: The stupidest thing we ever did was give up our 200-mile limit around Scotland. Now we've got every Tom, Dick and Harry hoovering up fish as if there's no tomorrow. Which there won't be. We've already got non-EC members out here. And in 1992 we'll have the Spanish as well, because they've joined. They've got, what is it? The world's fifth biggest fishing fleet?"

"It's not only the numbers," says Graham, "it's the technology. Progress is killing the fishing industry. Nowadays you can spot a sardine at twenty miles, shoot your net to almost any depth, sweep the sea bare. And nobody has the will to stop it. Take mesh size as a single example. We're restricted to a ninety-millimeter mesh. That's fine. We've tried all sorts of sizes and combinations in the past and we've proved you can keep fish stocks up if you use a ninety-mill. diamond with an eighty-mill. square panel. Imagine your trawl net, right? Like a great sock. It's all ninety-mill. diamond except for a strip around the top of the ankle. That's eighty-mill. square. Square mesh doesn't close up when you put a tension on it. Now, when your fish see the headline going overhead they're already in the mouth of the net. Their reaction is to swim upward. If there's a square mesh panel the small ones escape. They swim right through it while the bigger ones get swept on down to the codend. It's called a codend but it's got nothing to do with cod, ken? It means a sort of bag."

"As in codpiece?"

Graham says he doesn't know about that. "Anyhow, we have to use ninety-mill. diamond, although they're now talking about putting it up to 110, which will catch bugger all. If they do that, we're out of business overnight. They're always talking about conservation. Conservation this and conservation that. Well, we've proved you can fish with ninety-mill. and still have conservation. So guess what they've just told the prawners they can use? *Seventy* mill. and *two* nets per boat."

"Crazy, isn't it?" Donald asks the rev counter. The Kelvin diesel below vibrates reassuringly and jars the surface of his tea into a shimmer of concentric rings. "It's the truth what Graham's saying. Two seventy-mill. nets. Of course you need a smaller mesh to catch prawns, but since they're allowed to land a percentage of fish together with the prawns they just shoot their nets anywhere and take up every last tiddler. So much for conservation. And how is it conservation to allow the Danes to trawl for sand eels off Lerwick? They're only ground up for fish meal and animal food. Hundreds of tons of them, just to feed dogs and throw on the fields. You're not going to tell me it doesn't have an effect. There are half the seabirds here compared with a few years back. They lived on sand eels, you ken. It's down to greed, simple as that. Short-term profits today and hang tomorrow. What we've got up here is an entire industry in a mad scramble to cut its own throat."

A new sun below the horizon is beginning to disclose a haggard sea. As the dawn light strengthens, the surface takes on rumplings like a sheet of thin metal being shaken soundlessly. Out of pinkish, opalescent air the first fulmars arrive as if they knew the *Garefowl* would soon be hauling up. Shortly afterward she is hove to and wallowing in the choppy swell. The winch growls and wet hawsers begin sighing through sheaves, spraying drops of water. After some minutes the skulls bob up far astern and the cloud of fulmars circles and lands, coming ever closer. The heavy doors come up and are secured outboard, one on either side of the stern. They are freezing cold to the touch, black except along their bottom edges where they have been freshly scratched and polished by dragging along the seabed for four hours. Their cracks and joins are caulked with mud.

Hauling up is an immemorial business. The nets and skills and patience have all been deployed and now it is time to find out how well one's family will live for the next few days. Countless representations of the Sea of Galilee underwrite this moment: images of straining arms and bulging nets, of the glittering harvest of the deep. Nowadays powerful winches do most of the work, but there is still a fair amount of manhandling, of lumps of machinery with great inertia swinging dangerously while men dodge around the edge of the stern above a frigid sea. When the codend is finally hauled out the picture is suddenly not at all immemorial but dismally contemporary. A soft sack of fish appears, about the size of a bundle of hotel laundry and festooned with rubbish. The heads of flatfish stick out at all angles, eyes bulging with pressure, and from them and between them hang rags of plastic, trash bags, torn freezer packs, lengths of electric cable, flattened orange juice containers. The brutal meeting of still-flapping bodies with machinery is contemporary, too. As the codend is hauled free of the water the fish caught farther up the net are already being minced as they are dragged over the hydraulic pulley high above the stern. Their shredded bodies drop into the sea to be pounced on by gannets.

The full squalor of the net's contents is not revealed until the codend is swung inboard and emptied into a wooden sty. Into this mass of bodies smeared with gray North Sea silt wade men in Wellingtons, crunching and kicking among the dying, sorting out the unwanted with their feet: paint cans, a length of rusty chain, a battered steel drum, a work gauntlet, two beer bottles encrusted with growths, lumps of torn starfish, clods of jelly, the silvery sack of vacuum-packed coffee with the name of a Hamburg supermarket still legible. The rubbish of a thousand fishing boats and oil rigs and supply vessels is daily fished up, winnowed out and thrown straight back into the sea, building up on the bottom into an ever more concentrated and handpicked stratum of garbage.

There is about the architecture and layout of certain new housing estates on the edges of provincial towns something which makes it easy to guess they have been built on landfill over what, until a year or two ago, was the municipal dump. A clutch of abandoned gravel pits has been steadily filled with a million soup cans, dead refrigerators, burst sofas, outdated phone books and Dumpster loads of garbage-can contents. For years it was picked over by flocks of seagulls following a lone bulldozer as it leveled the heaps of rubbish. Finally, it was tamped down and topped off with a layer of soil from an excavation elsewhere in the county, and streets and houses and lawns were planted over the landscaped charnel. One imagines the street lamps as long hollow spikes driven deep into the festering seam to flare off the methane. It is a familiar inland prospect at the blurred borderland between the suburban and the subrural, where the terrain is anything but pristine.

Far off the coast of northeast Scotland at dawn, however, the view appears untouched. The blank light lifts like an ancient eyelid disclosing the glint of curved ocean in its timeless gaze. Gannets furl themselves and dive hard into the black-green surface; they are visible many feet below, streamlined as tenpins. In the cold, taintless air there is this same arctic clarity and there comes the memory of last night's weak flames and tattered banners draped across the northern heavens, the flickerings of an aborted aurora. To look into this grand expanse of movement and steely color is still to see what the Vikings saw, maybe, only for a moment, the idea of North; a vision which bursts as the eye catches on the horizon a structure hardening in the light—part Eiffel Tower and part Christmas tree. This single glimpse of an oil rig pricks and deflates everything. All too obviously it is no ship, not floating but *standing*. In the instant of this perception the North Sea sinks as though a plug has been pulled. It is no profound, pure abyss after all, but a lake shallow enough for man-made girders to stand on the bottom and poke through. Then also comes a memory of something Graham has said, that to trawl along the various pipelines for the different species which favor them as habitats (oil pipes are warm, gas pipes cold) has the disadvantage of filling the net with a mass of junk and hardware discarded by the men who laid the pipes and who regularly maintain them. Their courses are marked by tons of assorted soft-drink cans, plastic sacks of jettisoned nuts and bolts, paint drums still half full of toxic sludge, hundreds of fathoms of steel hawsers, sunken potato-chip packets.

One night (Donald had interrupted) his brother-in-law out of Peterhead was hauling up and when the codend swung into the arc lights above the counter every man aboard froze where he stood. Crushed into the meshes was the face of a girl looking out at them, her mouth open in a yell, her eyes wide. Partially lost among the plaice and whiting and dogfish were her twisted limbs. When the catch was released nobody wanted to wade into the bin to dig her out from beneath the bottles and squid and halibut, not least because there was movement everywhere as if things were trying to struggle up from the bottom of the heap. Eventually some brave soul pulled her out from beneath a heaving monkfish: a torn and deflated life-sized sex doll. Inside her mouth, molded into a red-rimmed O of insatiable accommodation, were hermit crabs. She, too, went back over the side, twentieth-century mermaid, Jenny Haniver herself, probably modeled from the by-products of the very same North Sea oil her roustabout lover had been helping to extract.

Now, on the shelter deck whose hatch is closed against the wind, Graham in Wellingtons crunches and kicks the dying, scooping up fish fourteen pounds or so at a time onto a sorting tray where Ian and Gordon quickly work through the catch. Into one plastic box they throw angler fish, into another cod, into a third plaice and sole. Then with a sweep the tray is cleared and down a stainless-steel gutter leading out over the scuppers shoot all the unwanted creatures together with bottles, plastic, bits of wire and shells. Fish of all kinds, wriggling and still, a hundred meals chucked back over the side in a slurry. Outside the gulls dive but pick and choose. The gannets prefer whole fish while the fulmars are really waiting for the gutting to begin. Behind the *Garefowl* stretches a sinking slick of bodies, for even the ones still wriggling will die. The violence of their capture and the hauling up through sixty fathoms has ruptured capillaries, made flotation sacs bulge out of mouths. Those fish with patches of scales torn off will anyway become prey to disease, to worms and saprophytes. One way and another they all drift back down to the lifting, swaying layer of refuse.

Any indignation at such "discards" is not the conventional "Think how many Third World families they would feed," since the same can be said at any other point in the chain of Western food production and consumption, from the processing of frozen vegetables to expense-account luncheons. The instinctive rebellion is more against the blithe inefficiency of a system whose regulations make it mandatory to junk a great percentage of what it catches; regulations, moreover, designed expressly to protect immature fish and replenish stocks. How powerful is this reversal of a famous miracle. Once, two small fishes allegedly fed 5,000 people. Our achievement is to make 5,000 fish feed two.

It is utterly violent, this daily wrenching of indiscriminate tons of living creatures up into the air, picking out those which happen to meet current laws and requirements, and returning tons of unwanted corpses to the sea floor. The sea floor itself has meanwhile been scoured by the trawl doors, its ecology (already laboring beneath a top-dressing of litter) further battered by the wholesale disruption of colonies of species that have nothing to do with human gastronomy and everything to do with the marine food chains: starfish, crustacea, weeds, corallines, urchins, jellyfish, algae. In addition there are the eggs of fish species such as the herring which are laid demersally, adhering to the seabed. It is never directly seen, this massive damage caused by the corridors scraped across the floor of the sea twenty-four hours a day, every day of the year, all over what were once the richest fishing grounds in Europe. (Silent beneath their dank quilt of freezer bags lie the lost birch forests of far-off Dogger.) What is seen in the codends of the nets is a representative sample, as the rubble in a Dumpster is the microcosm of a building site. I wander about the *Garefowl*'s stern, picking up dabs and other flatfish which have fallen to the deck and lie like dead leaves behind stanchions and around the bases of hydraulic gear, throwing them to the gannets. Maybe the only way to justify these creatures would be to be forced to eat them all, starfish included.

Over the next couple of days the boat sails back and forth, shooting and hauling up, shooting and hauling up. Twice the net comes up torn after a menacing concussion has thrummed through the boat. Each time Donald has hurriedly shut off the engine, for it means the trawl has hit a rock or some other obstacle. The North Sea is littered with wrecks. While Donald has marked many of them on his rolls of Decca Navigator traces there are always unmarked ones or fresh hazards.

"Bastards," says Ian as the trawl comes up with the coils of a hawser snagging the headline. "That's a towing cable from another trawler. Not lost, ken, just dumped. You've got two wire towing cables, right? About three or four thousand quid the pair of them. On average they'll last ten months. There's no scrap market for the old ones. You're forced to dump them. We take ours to a place where there's a hard bottom, preferably near a wreck, so nobody will be trawling there. But if you're not a local and don't aim to come back you don't care where you chuck your cables, do you? Bastards."

Although the catch is poor—Donald says it will not cover the costs of this trip, let alone pay anyone aboard—spirits in the wheelhouse are not low. Rather, everybody is philosophical. Sometimes you do all right, sometimes not. That's fishing. And what can you expect when the fishery is at its last gasp? (The radar screen is bright with the dots of other vessels: two Danish industrial ships, a German beamer, two Belgian trawlers, a Frenchman and a Dutchman.) Instead, conversation turns to what bad seamen some of the younger folk are, how they hardly know anything about the sea, but technology helps cover up their ignorance. The automatic pilot makes navigation simple, everyone agrees, but it is also dangerous, since it lulls the inexperienced into not paying enough attention and is often the cause of collisions.

"My father would have died of shame at the thought of colliding with another boat. Nowadays it's a bit of a joke. Insurance, you see." Suddenly the voice of an old, tight Scottish community is heard as they reminisce about the conventions, taboos and superstitions which even today rule some fishermen's lives. "My father was ashamed to do plenty of things people do now without a thought. Other things he just *couldn't* do, like put to sea on a Sunday. And there were certain words you wouldn't mention in front of him, he was that superstitious. You wouldn't dare say 'salmon,' for instance; it brought awesome bad luck. You had to get round it, say 'the silver fish' or something like that. Nor 'minister.' Nor 'pig.'"

"Nor 'rabbit,' come to that," puts in Graham. "It was a 'four-footer.' And nobody ever wore green. As for bodies, they'd never touch a floater. Never."

Their superstitions must have put them in the strangest positions. Men who would not touch drowned bodies would man the lifeboat and Donald's father, who had been coxswain, could not swim.

"You mean he was a lifeboatman and couldn't *swim?*"

"That's right. No more than I can, and I've been coxswain myself."

It turns out that only one of the *Garefowl*'s crew can swim. Donald was once knocked overboard wearing thigh boots, which promptly filled and began to drag him down like stones. He managed to catch a rope in time. "It wasn't good," is his description of being pulled below. "By rights I shouldn't be talking to you."

It is hard to know what to make of all this. They assert that at most 10 percent of Fraserburgh's older fishermen can swim and only a few more of the younger. They say this is because the young ones have been on survival courses run by the oil companies, but a more likely reason would seem that nowadays schools take their pupils to heated swimming pools. Undoubtedly it is evidence of a fishing community's strange and fatalistic relationship with the sea. In Britain, as elsewhere, there is a long tradition of neither fishermen nor sailors being able to swim. This is disdainfully brave or plain damn silly; else it stems from a dark belief that once you have fallen into the sea's grasp you are done for *by rights*. Or did a Puritan dislike of undressing, together with these latitudes' naturally frigid waters, always make of the sea a medium which was quite properly fatal?[3]

---

[3]To redress any implication that it is nowadays only in places such as isolated Scots fishing communities that people's lives are still affected by superstition, it is worth remembering that at the end of April 1991 the nation's Driver and Vehicle Licensing Centre in Swansea decided to stop using the number 666 on registrations because motorists' belief in the figure's satanic

We turn for Fraserburgh with fourteen boxes of fish. We must have discarded at least the same amount, more if dropouts are included (the fish that are shed from the net as it is hauled up). Donald thinks that if instead of throwing away undersized fish trawlermen were obliged by law to box and land them, the extra trouble would be enough to force them to adopt larger mesh sizes. Privately, I think it a pity he and his mates are so conservative in their tastes they will not even keep something back for themselves. They could eat fresh fish at least once a day (in three days we have eaten fish for one meal out of a total of nine) and I try to convince them how delicious most of their rejects are. In vain. No amount of insistence will make them taste octopus, crab, squid, gurnard or several other species. Donald says, "You get about twelve stones [168 pounds] of octopus to a box. On average they'll fetch four or five quid a box, maximum ten. Ten quid for nearly eighty kilos of fish? Isn't worth the effort. Dump them. . . . *Eat* them? I'll never eat one of those things. They all go to France. Places like that."

Finally, they hardly seem like people gathering food at all; it is merely something that can be sold. It might be manganese nodules we are bringing in. They are heading home to plates of Aberdeen Angus in thick gravy with tatties and bashed neeps.

~

There are several very curious things about the present state of the North Sea fisheries, not the least being that it is nothing new, not even the platitudes so often repeated in small wheelhouses: "The fisherman is his own worst enemy" and "Progress is killing the industry." The second half of the nineteenth century saw mechanization applied to fishing, advances which brought steam power to boats and winches and helped preserve and distribute catches by means of refrigeration and the railways. As early as 1863 a Royal Commission inquired into whether the

---

connotations was leading to a significantly higher accident rate. According to a DVLC spokesman quoted in the London *Evening Standard* (7 May 1991), "you see the number 666 in front of you and it makes you feel nervous. And, because you feel nervous, you bump into him." Compare this with the episode of the house President Reagan and his wife, Nancy, bought in Bel Air for their retirement. Its address was originally 666, St. Cloud Road. They insisted the number be changed to 668. Of an American president's day-to-day movements and decisions being determined to a large extent by the advice of professional astrologers we say nothing.

spread of trawling was perhaps beginning to deplete the North Sea stocks and decided it was not. By the turn of the century there was no longer much doubt on the part of anyone involved, with the exception of the government itself. Several peers knew better, but they were sportsmen. Lord Onslow, who had studied the matter at first hand and had already noted with distaste that the trawl crushed fish to death while the driftnet strangled them, determined to push legislation through parliament which would at least restrain trawlers from wasteful fishing. He proposed:

1 closing certain areas, either on a seasonal basis
  or for some years at a stretch;
2 extending the three-mile territorial limit;
3 a compulsory increase in net mesh size;
4 prohibiting both the landing and sale of flat fish
  below a certain size.

His attempt failed. Meanwhile, a fisheries expert named Aflalo was documenting the destructiveness of North Sea trawlers.

It appears that one shrimp-trawl, working in the Mersey, took 10,000 baby plaice in a single haul, while another in the same district brought up over 250 small soles, 900 tiny dabs, nearly 300 unmarketable whiting, 18 little skate, and some hundreds of useless plaice. When to this is added the fact that all this waste fish was accompanied by only twenty quarts of shrimps, some estimate may be formed of the terrible destruction for which such agency is responsible.[4]

That was written nearly ninety years ago. He went on to put the entire problem in a way which could scarcely be bettered today. Having referred to North Sea practices as a "conspiracy of depletion," he recommends fish farming as "making it possible to sow as well as to reap" and concludes:

It is when man shall have discovered the means of restocking the sea and of controlling its supplies that his 'dominion over the fish' will be perfect. The power to deplete, which so far marks the utmost limit of his advance, is mere

[4]F. G. Aflalo, *The Sea-Fishing Industry of England and Wales* (1904), p. 56.

tyranny. Dominion should embrace a more benevolent sway, and to that end no doubt the efforts of science and the might of law will presently join forces. It is to be hoped that the present friendly collaboration of the Northern Powers in the great sea in which they have a common interest may be the basis of a lasting harmony, more durable than any evolved in Utopian deliberations at the Hague.[5]

Almost a century on, hardly a word of this needs to be changed except to substitute "Brussels" for "the Hague" and to note that fish farming now exists to a limited extent, but in areas such as fjords and estuaries annexed for this purpose. There is no attempt to restock free-for-all fishing grounds like those of the North Sea. Much of the problem is that science has not yet lived up to Aflalo's expectations and extraordinarily little is still known about the behavior of fish, of their breeding and feeding and migratory habits, of the influence on them of changes in current, temperature and plankton levels. As Donald sardonically remarked, "The government have no idea when cod and haddock spawn."

In 1902, two years before Aflalo made the above remarks, the International Council for the Exploration of the Sea was founded. It is this same organization, ICES, which today is responsible for making the scientific assessments of fish stocks on which the EC bases its quotas: the so-called TAC or Total Allowable Catch. (It should be noted that TACs apply to fish *landed*, not to fish *caught*. It is not illegal to catch and kill large numbers of immature fish.) From the year of its founding ICES became a focus for all aspects of biological oceanography and fisheries research. Put succinctly its remit, according to one ICES member, was "to tell the commercial world how far greed might safely go."[6]

ICES's first task was to discover why fish have a habit of coming back to the same grounds for decades at a stretch, long enough to generate a local fishing industry, and then one year vanish altogether. It was supposed that a change in the currents carrying plankton might be one explanation. Other ICES scientists tagged fish in order to study their movements. Still others concentrated on refining a method of dating fish which had originally been used to determine the ages of carp, that of counting the growth bands on a single scale. They noted that fish grew

[5] *Ibid.*, p. 375.
[6] D'Arcy W. Thompson, "The Voyages of the *Discovery*," *Nature* 140 (1937), p. 530.

faster in summer and hence the bands on their scales became wider. It was a Norwegian, Johan Hjort, who finally discovered that the dynamics of fish populations are quite unlike those of terrestrial animals. Any single haul might well contain fish from two to twenty years old, and in very uneven distributions. He invented a terminology, calling fish born in a particular year a "year class," then set about trying to explain why there were good and bad year classes. A good year class would dominate catches for several years at a stretch while members of a bad year class, even though they might have grown well as individuals, were thinly represented and for shorter periods. This was mystifying. Some unknown factor—not obviously the weather—evidently allowed great numbers of fish to survive one year but then killed off young "recruits" in successive years. (No less mysterious was Hjort's describing fish in terms of a military academy. This had no connection with "schools" of fish, either, which derives from the Dutch word for shoal.)

On the other side of the Atlantic, meanwhile, ICES had been anticipated by all of thirty years by the US Fish Commission. This was the brainchild of Spencer Fullerton Baird, who set up his base in a tiny village near Cape Cod named Woods Hole, which he knew and liked from having spent his holidays there. His laboratory, together with its research ship, named the *Albatross,* offered the best facilities in America for marine biology. Baird died, much lamented, in 1886, not before T. H. Huxley had praised him for his work on fisheries which had yet to be done in British waters:

If the people of Great Britain are going to deal seriously with the sea fisheries, . . . unless they put into the organization of the fisheries the energy, the ingenuity, the scientific knowledge and the professional skill which characterizes my friend Professor Baird . . . their efforts are not likely to come to much good. I do not think that any nation at the present time has comprehended the question of dealing with fish in so thorough, excellent and scientific a spirit as the United States.[7]

[7]Quoted in G. Brown Goode, *The Smithsonian Institution* (1897), p. 88. To judge from the way he was remembered years later, Baird was an exceptional man. In 1918 a friend, Edwin Linton, paid tribute to him in elegiac vein:

I remember the day and the hour. It was afternoon, and the tide was low. I recall a picture of a red sun hanging over Long Neck and reflected in the still waters of Great Harbor, of

Two basic questions which Hjort posed have still not been answered to complete satisfaction. One was: Which period of a hatchling's life is the most critical? and the other: What are the factors which most affect its health and development? It is now believed that a good year class results if there is a rich supply of plankton during the first ten days after the larvae hatch. In view of the advances over the last 100 years in science of all kinds, including biology itself, this tentative conclusion scarcely shows a great breakthrough in knowledge of the life cycle of fishes. This in turn makes it odder still that governments on both sides of the North Sea have done so little to curb the "conspiracy of depletion." After all, there has been overwhelming evidence of serious overfishing throughout this period, among the most convincing of which was the effect of World War I. Submarine warfare and minefields closed many fishing grounds for four years, allowing fish stocks to regenerate. With the end of hostilities in 1918 North Sea fishermen landed fabulous catches, not simply greater numbers of fish but bigger specimens. This bountiful state of affairs lasted a short while, then deteriorated again so that by the mid- to late twenties fish stocks were back down to where they had been before 1914. The obviousness of such evidence is never enough to convince governmental exponents of *laissez-faire* policies. Even in the middle of World War II, Dr. E. S. Russell, the director of fishery investigations at the Ministry of Agriculture and Fisheries, could in the same breath urge against overfishing during the postwar reconstruction and also think it "probable that surface fish, like herring and mackerel . . . although a great source of food and wealth, exist in stocks too large to be depleted by fishing."[8]

To put it bluntly, nobody has the remotest idea of the full long-term effects of continuous overfishing. It has recently been suggested that it might be causing genetic changes in fish stocks as the fish, fighting for survival, are pressured into earlier sexual maturation at the expense of

---

sodden masses of seaweed on the dripping piles and on the boulder-strewn shore; and there rises again the thought that kept recurring then, that the sea is very ancient, that it ebbed and flowed before man appeared on the planet, and will ebb and flow after he and his works have disappeared; and in a singular, indefinite impression, as if something had passed that was, in some fashion, great, mysterious and ancient, like the sea itself.

(From E. Linton, "The Man of Science and the Public," *Science* 48 [1918], p. 33).

[8] *The Times* (London), 10 March 1942.

growth.[9] Shore stocks may also be driven wholesale into deeper waters, as has happened in the Grand Banks fishing grounds off Newfoundland, with unknown consequences for the area's ecosystem. Least of all can anybody say what impact Russia's new and powerful krill-fishing industry will have on South Atlantic waters, on baleen whales and penguins or—come to that—the entire food chain of Antarctica. In the 1988 season more than a third of a million tons of krill were taken. By now, with a new fleet and the urgency of a national food shortage, a million or more tons per season are probably being harvested.[10]

~

One day in January 1988 two companions and I were in a skiff midway between two islands in the South China Sea when we spotted something curious in the water ahead. There was a ragged patch like floating seaweed or the top of a reef as the tide begins to fall. We knew the area well; we were far out in a deep channel where there were no reefs. Also puzzling were the antics of a small white tern which was standing, fluttering and fluttering its wings as if in some difficult feat of balancing on this patch. As we neared we could smell it before we finally identified it as a mass of rotting animals. We identified fish, a baby dolphin and various bird carcasses. The tern's desperate agitations increased with our approach and we could now see that its feet were caught in the nearly invisible meshes of a ghost net, a fragment of drift net which floats and continues to catch almost any animal that comes into contact with it. With some difficulty we freed the little tern, the bottom of the skiff whispering over nylon and fins and bumping softly among the heavier corpses. There was nothing else to be done. Sooner or later, the gases of putrefaction having been released, the mass would sink beneath its own weight until fresh gas was generated or its heavier contents broke up sufficiently to filter out between the meshes. Then it would rise to the surface once more.

Since Japan first began using nylon monofilament nets for drift netting in its own coastal waters in 1976, there has been a great expansion of pelagic drift netting. Over the last few years that in the Central and

[9]See Richard Law, "Fishing in Evolutionary Waters," *New Scientist* 129, (London), p.1758 (2 March 1991).

[10]See Paul Brown, *The Last Wilderness* (1991).

North Pacific areas has been the principal focus of public and government attention, tending to obscure the fact that this practice is widespread in coastal areas as well. In archipelagic seas it has caused immense
damage to marine and bird life and is now affecting the livelihoods of
local fishing communities. Pelagic fish are those which feed at or near the
surface, usually in large, dense shoals. To catch them, modern drift-
netters hang an invisible curtain of nylon mesh forty feet high in the sea
and leave it to drift with the tides, currents and winds. Drift netting is
a long-established technique, one which has been used for decades in
British inshore waters where to this day nets to a maximum length of 2.5
kilometers are perfectly legal. What is new about the practice since 1976
is the use of invisible and virtually indestructible nylon mesh.

For six months of the year the North Pacific is invaded by more than
1,500 ships, mainly from Japan, Taiwan and South Korea, together
forming the world's largest fishing fleet. Every night each ship lays
between thirty and fifty miles of net, leaving it to drift for six to eight
hours, so that in these latitudes on any night there can be anything up
to 50,000 miles of net deployed, about enough to encircle the Planet
twice. The ships are highly efficient, with the latest electronic means for
detecting shoals, especially those animals rising vertically each night as
the Deep Scattering Layer, and some with additional sonic aids for
herding them more tightly. The nets catch almost anything which
touches them, including birds and especially cetaceans, to whose sonar
the filaments are invisible. Marketable fish are kept; nontarget species
and those of no immediate economic interest are dumped overboard by
the ton. It is a system which has been trenchantly described as "strip-
mining the oceans."[11] From the fishermen's point of view it is cheap and
simple; from that of the ecosystem the effects cannot be determined
since so little is known. But its results are not invisible. Even Japanese
drift-netters now admit they see far fewer dolphins, seals, juvenile hump-
back whales and seabirds than only five years ago. Japan has long since
banned drift netting in its own territorial waters because of the threat of
ecological collapse.

The activities of this industry first became known as late as 1983, and

---

[11]A phrase coined by the Hawaiian-based Earthtrust group who, in the last eight years,
have painstakingly brought to light much of the information now known about this secretive
industry.

then only because Japan had formally requested US permission to go drift netting off Alaska within the 200-mile EEZ. The intention was to intercept Pacific salmon on the way to their river spawning grounds. The US granted the permit, believing that these stocks were heading for Russian rather than North American rivers. Later, when the Japanese tried to renew it, the questionnaire they had to submit unwisely revealed that a typical season's catch included an estimated 14,000 Dall porpoise as well as three quarters of a million seabirds. (At the time the Japanese salmon fleet comprised 172 vessels, each with nine miles of net, totaling a mere 1,548 miles as compared with that currently being deployed by the North Pacific red squid and tuna fleets.) As it happened, the salmon stocks were North American after all. This incident provided an excellent demonstration of how little is known about the movements of fish. The heirs of Spencer Fullerton Baird had misread the migration patterns of the most economically important species in Alaskan waters. This is to leave aside the ethics involved in the assumption that since they were Russian salmon they were fair game for plunder. It was only when the Alaskan and Canadian salmon fishermen began to suffer ruin from the depletion of their stocks that the error was acknowledged. In 1988 Japanese fishing boats were banned from US territorial waters, two years after a similar ban by Australia.

East Asian markets are regularly glutted with salmon still being caught illegally in Alaskan waters. This drift-net fleet probably takes 50,000 tons of salmon annually and Earthtrust has estimated that the number of "dropouts" and injured and dying "escapees" equals the same amount again. If this can happen within the United States' EEZ, it is not hard to imagine what goes on in international waters. There is a single Regulatory Zone in the entire Pacific, that for the red squid fisheries of the central region. The drift-net fleets driven out of national waters have congregated here but pay scant attention to rules in this vast, unpoliced area. Under international agreement the limits of this protective zone advance northward in monthly steps from June to September to take account of the breeding season, but boats are frequently seen fishing outside it.

From time to time a brief holiday is possible from Bertrand Russell's "unyielding despair." In late 1991, under fierce international pressure, Japan finally announced its suspension of all drift-net fishing by 1993, an action which should lead to a worldwide ban. Hats in the air, then.

But those with a sense of history will retain their pessimism as a thwarted billion-dollar industry develops instead the technology for precisely targeting species, with the danger of their eradication (as in the case of the majestic bluefin tuna). "Quotas" will once more be fixed by authoritative-sounding international bodies who cannot tell how much loss the ocean has already borne, nor guess what it still can bear. If the North Sea, surrounded as it is by a community of developed nations with access to the best scientific information, can be systematically ruined through conflicts of interest and political expediency, what real hope is there for the Pacific? Who is to police its vast, unclaimed areas, and to enforce what law? Under the virtuous flag of an abuse redressed, cynical attrition is as likely as ever to slither around laws and evade patrols, just as it does in the tiny pond of the North Sea. The issues will float up and down for years to come, like ghost nets with their rotting cargoes.

For now, even people with no immediate connection with fishing have noticed a decline in Pacific wildlife. As a geophysicist aboard *Farnella* said one afternoon, surveying from the rail the empty expanse of ocean on which floated a sheet of tar-stained plastic where eight or even five years ago a school of dolphins might have sported, "We don't deserve this world." The drifting refuse, the absence of any sign of life for two weeks but for a few flying fish and the occasional mournful bird, was bound to make anyone take stock. "*Homo sapiens sapiens.* In this single century we've slaughtered a thousand times more people than all the Genghis Khans of history put together. Into the bargain we've laid waste our planet. Not bad going for a mere hundred years. And look what we've got in return. Machines for mapping the ocean floor and a brave new race. We've had the Beaker Folk and now we've got the Consumer Folk. Safeway Man. *Homo supermercatus.*"

~

The trait in the human species of harvesting first and assessing the consequences at leisure is clearly a genetic inheritance. Unlike felines, which eat their fill and walk away, we are in this respect closer to the canines such as foxes, which kill an entire roost of chickens they will neither eat nor bury for later. There is something hopeless in *Homo*'s mixture of brutality and compassion, a cross-purpose of muddlement. Even as Spencer Fullerton Baird was founding Woods Hole, Wyville Thomson was trying to observe the creatures of the North Sea. He wrote of English fishing smacks being welled to supply fresh cod for the London market (this was before refrigeration). A large, square tank was built amidships with holes for fresh seawater to circulate. The fish, he noted, were oddly tame:

It is curious to see the great creatures moving gracefully about in the tank like goldfish in a glass globe. . . . They seem rather to like to be scratched, as they are greatly infested by *caligi*. . . . One of the fish had met with some slight injury which spoiled its market, and it made several trips in the well between London and Faëroe and became quite a pet. The sailors said it knew them. . . . It was always the first to come to the top for the chance of a crab or a bit of biscuit, and it rubbed its "head and shoulders" against my hand quite lovingly.[12]

Treating a codfish as another Englishman might treat his hound may have struck his contemporaries as eccentric, especially since they were accustomed to a reign over the animal kingdom, which was in general less than benevolent and often maniacal. Accounts of the huge eighteenth- and early-nineteenth-century massacres of creatures such as whales, seals and sea lions make painful reading; even at the time explorers and others sensed that something was not right. W.H.B. Webster, the surgeon aboard HMS *Chanticleer* during her voyage of 1828–30 in the South Atlantic, wrote:

The harvest of these seas has been so effectively reaped, that not a single fur-seal was seen by us, during our visit to the South Shetland group; and, although it is but a few years back since countless multitudes covered the shores, the ruthless spirit of barbarism slaughtered young and old alike, so as to destroy the race. Formerly two thousand skins a week could be procured by a vessel; now not a seal is to be seen.[13]

Captain James Weddell, making much the same journey five years earlier, had noted that both sea lions and fur seals were almost extinct on South Georgia, "not less than 1,200,000 skins" having been sent to the London market, together with 20,000 *tons* of "sea-elephant" oil. He evidently liked penguins, observing that "Sir John Narborough has

[12]C. Wyville Thomson, *The Depths of the Sea* (1874), p. 59.
[13]W.H.B. Webster, *Narrative of a Voyage . . .* (1834), vol. II, p. 302.

whimsically likened them to 'little children standing up in white aprons.' "[14]

When fur seals became scarce the "sea-elephant" (elephant seal) was turned to for its oil and Weddell noticed the animal's pitiful docility. "It is curious to remark, that the sea-elephant, when lying on the shore, and threatened with death, will often make no effort to escape into the water, but lie still and shed tears, merely raising the head to look at the assailant; and, though very timid, will wait with composure the club or lance which takes its life."[15] In order to extract the oil from the seals' blubber it was flensed and rendered down in vast iron caldrons. Fuel being extremely short in these latitudes the sealers burned penguins, each of which contained about a pint of oil. Finally, the supply of seals ran out and the sealers had to turn to the penguins themselves. "In common with all Antarctic creatures," observes Edwin Mickleburgh,

they had little or no fear of man and it was therefore a simple business for the sealers to drive them into pens where they were clubbed and flung into the bubbling pots. It is recorded that some birds, in the interests of faster production, were driven down crudely constructed gangways directly into the pots where they were boiled alive.[16]

Not that this sort of thing was confined solely to sealers and whalers. W.H.B. Webster, calling at Buenos Aires on his way south, had already heard a local resident describe how he had "once sold a flock of sheep, amounting to two thousand, at 1s 6d. [fourteen cents] per head, for the sole purpose of fuel for a brick-kiln."[17]

Looking once again at the North Sea and the Pacific Ocean at the end of the twentieth century, we can see that there is little new about the creation of semideserts where once teemed marine and bird life. Nor, in its capacity to kill indiscriminately, does modern drift-net fishing really mark a new departure. Probably the only novelty is that of pollution, of animals being sieved from the sea and replaced with bottles and sheets of tarry plastic. Even in this matter the same muddlement and cross-purposes can be found, the same ironies. Thus, scientific bases in Antarc-

[14]James Weddell, *A Voyage Towards the South Pole* (1825), p. 53.
[15]*Ibid.*, p. 136.
[16]Edwin Mickleburgh, *Beyond the Frozen Sea* (1987), p. 31.
[17]Webster, *op. cit.*, vol. I, p. 83.

tica, set up precisely to study a pristine, nearly sterile environment, have themselves polluted their surroundings. The worst offender has been the US base at McMurdo Sound, where surplus bulldozers have been casually tipped into the sea and open-pit burning and landfilling of wastes have been everyday practices for years. Efforts are continuing to clean it up and many another base, newly uncomplacent, is discovering how expensive and inconvenient it is having to ship out hundreds of barrels of frozen feces and urine in addition to the other refuse it generates. There is an irresistible parallel to be drawn between Antarctica and another pristine—and wholly sterile—environment: space. Since the space age started in 1957, 4,000 satellites have been put into orbit, of which some have fallen but plenty remain. In addition to these and their defunct rockets there are an estimated 70,000 fragments of junk, including screwdrivers. If in the next three decades the amount of debris in low Earth orbit increases at the same rate, it will total three million tons and render any further human space exploration prohibitively dangerous. Various hugely expensive schemes for cleaning up space are currently being considered.

~

Until recently I was finding it easy to compare modern fisheries such as those in the North Sea with what it pleased me to think of as "true hunting." Because I spent time each year hunting my own food in the sea I deployed sentimentalities in my own favor. Certainly, when one takes on individual fish with a home-made, elastic-powered speargun of very limited range and accuracy, using only the air in one's lungs, the odds are heavily in the fish's favor. No doubt also it attunes one to things previously obscure. In the lengthy daily process of stalking food one learns much about different species of fish and their behavior. One also becomes observant of other phenomena, noticing weeds, currents, sudden thermoclines, coralline animals and the local benthos, light and shade and color. One also discovers things about one's own body: how to control breathing and how to lie at different depths, as well as the graph which plots time spent in the water against increasing frequency of urination, the remarkably dehydrating effects of three hours' constant diving and underwater swimming. There are unlikable character traits to confront, too, including those pairs of seeming opposites: callousness and fear, impatience and hesitancy.

There is one particular sentimentalism about hunting which all of a sudden I do not like at all, maybe because I once partly subscribed to it myself. It is that which speaks of a deep, quasi-mystical "understanding" between hunter and prey: a sort of mutual respect where after hours of effort the hunter is half pleased when his quarry escapes or, conversely, when it seems almost content to die. Presumably this derives from the humbug of chivalry and the codes of jousting. The thing in hunting is to win. When the novelty of the experience has worn off and the basic technique has been acquired, there remains the task of getting one's food as quickly and efficiently as possible because there are plenty of other things to do, such as collecting firewood, making another spear, repairing a rickety fish drier or just sitting under a shady tree. At this level hunting is simply gathering food, a necessary and often pleasurable chore. It is quite different from those grand, allegorical duels between old men and the sea, or grizzled captains and white whales. Yet to fish day by day off the same stretch of shore and, where there is a long fringing reef, off the same groups of corals, is to see from within the impact of local fishing. Nowadays I prefer to swim out beyond the reef, to go out at night and instead of killing parrotfish in their holes or mullet asleep on the sand wait for the solider pelagic species such as pampano to come in. At the extreme edge of the flashlight's beam a pale shape is glimpsed for a second. It might have been imagination or else a momentary fault in one's retinal wiring (pressure does strange things to night vision). It is worth pursuing, though, and one holds the beam on the spot where the object may have been and makes a burst of speed with both plywood flippers. At night most fish are either immobilized by darkness or else vanish with a fin's flick; pampano are strange in that they seem to allow themselves to be pursued, partly alarmed and partly attracted by the light. They could easily escape, but often after an exhausting chase one can overhaul them. In the light they are round and silver, about the size of a dinner plate, and in contrast to most coral species there is good meat on them. The technique is to hold the flashlight at arm's length and out to one side. Like all laterally flattened fish pampano turn so as to present an attacker with an edge-on view, but this attacker has outthought it and for a moment it is almost sideways on. If the aim is true and the speargun works, that moment should be enough. The range is never more than seven or eight feet. More than that and the spear will not penetrate. Twice that distance is the effective limit of visibility with a two-battery flashlight. A brace of pampano (for if there is one, there will be others) is plenty. On a good night one may be out and back and building a fire within forty minutes.

In much of Southeast Asia the pressures of virtually unregulated commercial fishing have led to close parallels between the fishermen of villages like "Sabay" and those of Fraserburgh. In Scotland it is—incredibly—not illegal to trawl right up to the beach. This practice has done enormous harm to littoral fish stocks, as to the locals who in calm weather could once go out in small boats and fish safely within a few hundred yards of shore. In the Philippines the equivalent is provided by the proliferation of *buli-buli* and *basnig* fishing. *Buli-buli* refers to large seine nets of very fine monofilament mesh, often as small as ten millimeters. *Basnig* uses similar nets but at night, with banks of bright lights to attract squid and nocturnal species. The older craft use propane gas for their lamps, the modern ones electricity. A *basnig* fleet with its cityscape of lights and distant massed chugging of generator engines is a characteristic sight. From its deployment one can often tell as much about local politics as about the offshore reef formations keeping it at a distance. Officially, there is a seven-kilometer limit inside which only "sustenance fishermen" may legally fish. In addition, Fisheries Administrative Order no. 164 places restrictions on all *buli-buli* fishing. It is illegal to use a boat of more than three gross tons, as it is to use a net whose stretched mesh is less than 29.9 millimeters. Smaller boats using legal nets may fish within the seven-kilometer limit but come into the jurisdiction of the local municipality. Violent battles sometimes erupt between *basnig* and *buli-buli* fishermen on the one hand and locals on other who claim their livelihoods are being ruined. They are not wrong; inshore fish stocks are visibly depleted after a single night's close approach by illegal fishermen.

As in Scotland, there is the same acknowledgment that unpoliced practices combined with overfishing can drive small fishermen to crime in order to stay alive. In many provinces there is a steady battle between thinly stretched, and sometimes conniving, local authorities and an army of people who go fishing with homemade dynamite, cyanide and bleach. In fresh water, electricity is also used. One can occasionally see small men staggering about in a shallow river with a twelve-volt Jeep battery strapped to their backs, prodding the water with two terminals. Poison is mostly used to stun reef species for export as aquarium fish. Those it does not kill outright it weakens, and it is estimated that maybe three quarters die within a fortnight. It also kills coral polyps. So does dynamite; and the skill with which it is often both made and used does not alter the fact that nontarget species die as well, and in deeper water, retrieval rates may be fairly low.[18] But as local fishermen—who are neither blind nor stupid—will say, What can you do when you have a family to feed and fish are so scarce?

In all this anarchy there is one thing to be said about fishing at local level in Southeast Asia: Precious little goes to waste. There is hardly a species which is not eaten, nearly nothing too small to eat. In the case of fried fish, much of the skeleton is often eaten as well. Even the tiny conical hearts of certain mackerel-like *Scombridae* are saved. To this extent local fishing in the developing world is free of the cavalier squandering and disdain which accompanies commercial fishing by the wealthy nations. To watch the fishermen of "Sabay" and their families handling fish, whether alive or dead, is to witness a radical respect for food.

At the extreme opposite is the deep-frozen brick of supermarket cod, prawns, sole fillets, tuna steaks. This sanitized object represents merely the pinnacle of an industrial pyramid of slaughter, destruction and waste. To speak with refined, Western sensibilities in mind: In terms of seemliness it is no longer possible to propose fish-eating as somehow less objectionable than meat-eating. In terms of ecological damage, the worldwide plundering of marine life may turn out to have been even more disastrous than the felling of rain forest for the benefit of beef ranchers.

[18]For a fuller description of these methods see James Hamilton-Paterson, *Playing with Water* (1987).

Popular conservation has helped promote the notion that there must have been an ideal "balance of Nature" which existed before *Homo* began upsetting it. This is nonsense, of course. The history of life on Earth is full of episodes like the great Permian extinctions at the end of the Paleozoic. These not only wiped out the ubiquitous trilobite, which had survived since the Cambrian, but about 96 percent of all living species of fauna as well.

If there is an unfortunate consequence of James Lovelock's elegant and serious "Gaia" thesis, it is that it lends itself to being hijacked by all manner of fringe theorists and used to support their own dotty notions. Thus, the idea that the biosphere might in some sense be self-regulating so as to maintain conditions favorable to life has been perverted, turning the Planet into a sentient Mother Earth. Anthropomorphizing it as "she" makes "her" struggles to maintain an immemorial balance in the face of *Homo*'s despoliation seem intentionally remedial, even noble. In this reading "Gaia" comes across as a somewhat sainted landlady, trying her utmost to accommodate her latest lodgers, who have turned out to be slobs and vandals intent on ruining her delicious mansion. Such an idea is absurd and unscientific, and should not be blamed on James Lovelock. It is what happens when people full of millennial *Angst*, guilt feelings and moralized views of biology adapt a serious hypothesis for their own purposes. They are dealing with a goddess, or maybe an old tyrant who once let all but 4 percent of her animal lodgers die out. We are dealing with evolution. Conditions on this Planet are constantly changing, and always have been. The extinction and evolution of species is as mutually dependent and constant a flux as the Earth's crust is itself plastic, wobbling to the belches of vulcanism and to the tuggings of celestial bodies. *Homo sapiens sapiens* is only one of an estimated thirty million species, and it is possible to argue that since he is as "natural" a creature as, say, the crown-of-thorns starfish, *Acanthaster planci,* any results of his presence, no matter how devastating, are also "natural."

There is often a blurring of motivation in public "Green" campaigning which amounts to a principled dishonesty. On the brink of the year 2000 the perception is that *Homo*'s remodeling of the balance of the biosphere has fatally endangered everything. The astutely televisual selection of species of cetaceans and large mammals (whales, dolphins, elephants, pandas) has actually come to stand for the sudden awareness of *Homo*'s threat to *himself.* In the last ten years there has been a panicky rise in the number of TV and radio programs, magazine articles and scientific papers devoted to the physiological and psychological threats to man's continuing existence. In the first case it is constantly repeated how menaced he is by his own poisons, effluents and by-products, as well as by his thoughtless exploitation of delicate global structures. In the second, it is a darkly lurking shibboleth that urbanization leads inexorably to a breakdown of social behavior, to mental illness, endemic envy and dissatisfaction, to out-and-out psychosis and mass murder ("Just look at what's happening in America . . ."). Man is perceived as being threatened simultaneously from without and within. The timely adoption of lovable big animal species to gather like a lens the varied gazes of his self-concern cannot be dissociated from the approaching millennium. We reach for pandas as a child for his teddy.

It is entirely proper to wish to preserve every species on Earth, whether (to our eyes) magnificent creatures or humble slime-molds, and not only because of their interdependence. *Homo* is an aesthetic animal and has pungent notions of the sublime. He feels diminution when the familiar vanishes. He experiences with the greatest upset the ghost of loss which stalks his waking days as well as his sleeping hours. It mobilizes in him a tenderness akin to vulnerability, to the point where a large part of his wistful concern for whales and the environment generally is displaced fear for himself, an intense longing looking for somewhere to alight. It becomes vital for him to know that whales and wilderness still exist somewhere on Earth, even if he never sees or experiences them at first hand. They represent cardinal points on the map he has inherited from his ancestors, representations of an earlier world where such things fixed the terms of daily living. Obscurely he feels that without them he is lost, or at least that he cannot shed them without consequences he can neither foresee nor articulate. If only he would say so! The high principles are there, to a fault; but so is dishonesty, because in shifting attention onto certain choice species he postpones recognizing his poignant lack, his own strange delicacy. To this extent *Homo* needs more self-interest and not less, provided it attends to that interior chart, to the space which is the ancient and common legacy of his species and whose territory he necessarily inhabits.

What alarms him most and makes him most confused and despondent is the speed at which environmental changes have taken place. This is a comparatively new phenomenon. It is not until the late nineteenth century that a sustained note of mourning is heard for parts of the familiar scenery that have vanished. Since it had to do with a rapid rise in industrialization and population, this note was first heard in Britain. It was typically sounded by Gerard Manley Hopkins in his 1879 poem "Binsey Poplars," on the cutting down of a row of trees near Oxford to make way for housing ("After-comers cannot guess the beauty been"). Thereafter it is heard increasingly, quite distinct in tone from a generalized literary *eheu fugaces*. The snows of yesteryear are, after all, nearly always followed by the snows of this year and those of next. Hopkins was recognizing something new: the destruction of trees that never would be replanted.

Since then, this process has accelerated so that today only the most purblind twenty-year-old could fail to notice a vanishing of things even since his own childhood. Those twice his age and more may have the vividest childhood memories of entire landscapes which are now gone, while people brought up in horsedrawn days before World War I can find themselves in alien terrain. This is a disturbing thing to have happened and leads to the ironic conclusion that it may often be easier to deal psychically with sudden disaster than with steady attrition. French and Belgian farmers whose fields were churned into wastelands by trench warfare knew they had only to wait a few years before most of the greenery sprouted back again. Conversely, those people who grew up to the rich, tangled hedgerows and small fields of the British countryside in the thirties, forties and even fifties know they will never see them again, nor many of the flora and fauna associated with them. The process is mostly irreversible, not because the species would not return if allowed to or—like a row of poplars—were replanted, but because roads and housing are never plowed up but only proliferate.

Looking back, it strikes us as curious how seemingly unaffected people once were by the extinction of a species within their lifetime. That was before such things took on the ominousness of the commonplace, as though they were isolated zoological events without particular consequence and foreboding nothing. Only now and then is it possible to detect misgivings beneath the breezy pragmatism, even sometimes a current of unease. "Last chance to see what Audubon saw," wrote an unnamed columnist standing in front of a cage in Cincinnati in 1913. This contained "Martha," the sole remaining passenger pigeon in the world. This species, *Ectopistes migratorius*, had within living memory been the most familiar bird in the whole of North America and was prettily illustrated by Audubon in his *Birds of America*. Its wholesale slaughter by settlers peaked in the mid-nineteenth century with massacres involving entire flocks of tens of thousands of birds. The last flock of this handsome creature—it was somewhat larger than an English woodpigeon and beautifully marked—was sighted in Illinois in 1895. The last recorded wild specimen was shot in Quebec in 1907. "Martha," meanwhile, had been bred in the Cincinnati Zoo in 1885. She died at 1:00 P.M. on 1 September 1914, aged twenty-nine, and with her an entire species. "The main cause of their extinction is plain for all to see," wrote Richard Carrington some forty years afterward. "Men cannot escape the moral responsibility for the callousness, the greed and the supreme irreverence for life that led to the passing of the passenger pigeon."[19]

This particular bird had followed many other species, which included the dodo—hunted to extinction in Mauritius in the late seventeenth century—and the great auk. This last, the garefowl, was also flightless. It was hunted in the North Atlantic for its fat, sometimes being killed because it was associated with superstitious belief. One was burned in Ireland in 1834 for being a witch. The very last specimen was killed ten years later. Today in New Zealand the peculiar kakapo is down to the last forty-odd specimens. This large green-and-yellow bird, something between an owl and a parrot, cannot fly, never attacks and never defends itself. When it can be encountered at all (for, scarcity aside, it relies on exquisite camouflage), the kakapo has a further claim to be remembered as an example of Nature's diversity, for it smells strongly of freesias. Aesthetics apart, the loss of a species has consequences which can never be entirely predicted. The extinction of the dodo eventually had effects on the very landscape of Mauritius, certain hardwood trees failing to seed themselves and beginning to die out. Only in recent years has it been established that the dodo played a central role in the seeding and distribution of these trees by eating their fruit. The bird's digestive tract softened the seeds' testa and assisted germination. Now turkeys are being imported from America in the hopes that they may act as substitute dodos.

While endangered species embody poignant reminders of our own mortality, it is the vanishing of entire landscapes that upsets us most. There is nowhere left to turn for solace and with which to re-create the continuity of our lives. Sights, smells and sounds may all vanish. I pretend not to mourn the wild profusion of the natty yellow-and-black striped caterpillars of the cinnabar moth which once stripped to the bone the clumps of groundsel to be found on every patch of wasteland in southern Britain. Likewise I miss the sheer variety of other butterflies and moths (including many rare species) which appeared even in the most suburban garden as late as the early sixties. One knew where Yellow and Red Underwings would be, and when in May to look for the Angle Shades moth just after it had hatched and its colors were at their freshest.

[19]Richard Carrington, *Mermaids and Mastodons* (1957).

The subtlest peach and brown and olive tints seemed to hover a fraction above the surface of its wings as if pure color stood off its scales by the thickness of dust, glowing and velvety. It now seems both important and hopeless to wish for other people such pleasure and ravishment, whether of looking at moths or being frisked around by dolphins. It is true that after-comers can never know exactly what they have missed; but missing things in our own lifetimes sets in motion the inarticulate hollowings of loss, and in turn we apprehend how quickly ordinary beauty is being made to vanish as if the hand of man held a wand whose touch made some things disappear for good and turned all the rest to lead. Each generation adapts to an impoverished world, but for the first time people are conscious of having to make do with remains. This has its effects.

In 1918 a steamer was wrecked near Lord Howe Island, about 400 miles off the east coast of Australia. Its rats swam ashore and changed the landscape forever. Two years afterward one of the islanders wrote, "This paradise of birds has become a wilderness, and the quietness of death reins where all was melody."[20] Such events have become common on islands, where casually or accidentally introduced domestic animals (goats, rabbits, cats and dogs) as well as vermin can decimate native species and upset an entire ecology.

Of all the laments for a vanished landscape, perhaps that by Edmund Gosse tells Britons most vividly what they have done and what they may no longer see. No chance shipwreck caused the disaster. On the contrary, if anything it was caused by zoology allied to a strong religious bent. At the ages of nine and ten Gosse spent countless hours with his famous naturalist father examining rock pools on the Devonshire coast, "a middle-aged man and a funny little boy." His account was published in 1907 and refers to the years 1858 and 1859, immediately after the publication of *Omphalos*. In the intervening half-century the whole seascape changed. The passage is a beautiful description of a lost world, and one of the first examples of a sensitive and passible man fully aware of what has gone and how its going was part of a pattern destined to be repeated at ever-gathering speed. In its prescience, it is a warning of the possible consequences of scientific fieldwork, even of "eco-tourism."

[20]Quoted in Rachel Carson, *The Sea Around Us* (1915).

If anyone goes down to those shores now, if man or boy seeks to follow in our traces, let him realize at once, before he takes the trouble to roll up his sleeves, that his zeal will end in labor lost. There is nothing, now, where in our days there was so much. Then the rocks between tide and tide were submarine gardens of a beauty that seemed often to be fabulous, and was positively delusive, since, if we delicately lifted the weed-curtains of a windless pool, though we might for a moment see its sides and floor paven with living blossoms, ivory-white, rosy-red, orange and amethyst, yet all that panoply would melt away, furled into the hollow rock, if we so much as dropped a pebble in to disturb the magic dream.

Half a century ago, in many parts of the coast of Devonshire and Cornwall, where the limestone at the water's edge is wrought into crevices and hollows, the tide-line was, like Keats' Grecian vase, 'a still unravished bride of quietness'. These cups and basins were always full, whether the tide was high or low, and the only way in which they were affected was that twice in the twenty-four hours they were replenished by cold streams from the great sea, and then twice were left brimming to be vivified by the temperate movement of the upper air. They were living flower-beds, so exquisite in their perfection, that my Father, in spite of his scientific requirements, used not seldom to pause before he began to rifle them, ejaculating that it was indeed a pity to disturb such congregated beauty. The antiquity of these rock-pools, and the infinite succession of the soft and radiant forms, sea anemones, seaweeds, shells, fishes, which had inhabited them, undisturbed since the creation of the world, used to occupy my Father's fancy. We burst in, he used to say, where no hand had ever thought of intruding before; and if the Garden of Eden had been situate in Devonshire, Adam and Eve, stepping lightly down to bathe in the rainbow-colored spray, would have seen the identical sights that we now saw,—the great prawns gliding like transparent launches, anthea waving in the twilight its thick white waxen tentacles, and the fronds of the dulse faintly streaming on the water, like huge red banners in some reverted atmosphere.

All this is long over, and done with. The ring of living beauty drawn about our shores was a very thin and fragile one. It had existed all those centuries solely in consequence of the indifference, the blissful ignorance of man. These rock basins, fringed by corallines, filled with still water almost as pellucid as the upper air itself, thronged with beautiful sensitive forms of life,—they exist no longer, they are all profaned, and emptied, and vulgarized. An army of 'collectors' has passed over them, and ravaged every corner of them. The fairy paradise has been violated, the exquisite product of centuries of natural selection has been crushed under the rough paw of well-meaning, idle-minded curiosity. That my Father, himself so reverent, so conservative, had by the popularity of his books acquired the direct responsibility for a calamity that he had never anticipated became clear enough to himself before many years had passed, and cost him great chagrin. No one will see again on the shore of England what I saw in my early childhood, the submarine vision of dark rocks, speckled and starred with an infinite variety of color, and streamed over by silken flags of royal crimson and purple.[21]

In its way this is a classic threnody, indulgently melancholic even to the furtive pleasure of "No one will see . . . what I saw." It also sets out the biblical coordinates of a mode which survives to this day as a shadow between the earnestly secular lines of much conservationist rhetoric: "the Garden of Eden," "paradise," "vision," "profaned" and, of course, the Father. This is touching in its confusion between the child who could not always distinguish his father from God, and the adult writing a devastating biography who could see the difference all too clearly. Both personae are there in this passage, unconsciously revealed by Gosse the son in two scientifically conflicting phrases: "undisturbed since the creation of the world" and "centuries of natural selection." The first represents his father's firm belief (he was a devout creationist), while the second is entirely his own, convinced Darwinian that he is. This identification of his father with the Creator, and the *ur*-landscape of his childhood in Devonshire with Eden, maybe gives an additional clue as to why we can feel so devastated by the disappearance of the place where we grew up. Landscape blurs easily into the parental.

This mourning for landscape, this apprehension of death without a proper body to grieve over, is one of this century's cruelest legacies. It has often been recorded, but seems to have gone largely unremarked as a likely cause of common forms of despair and depression. For lack of any medically plausible origin these are presumably attributed to the usual domestic disorders, disappointments and jiltings; whereas in reality the sorrow may be far grander, more pervasive and unsolaced, its cause misunderstood by both sufferer and doctor.

[21]Edmund Gosse, *Father and Son* (1907).

It is said that the British reading public's nostalgia for the imagined certitudes of Victorian and Edwardian England accounts for the popularity of rural diaries and reminiscences. This may be so; but it is quite as likely that the nostalgia is as much for a lost landscape as for any vanished social order.[22]

[22]Sooner or later this ailment will doubtless be officially recognized and accorded its own name—probably an inflated tag along the lines of :"Topological Mourning Syndrome," or even "TMS." Actually, the condition itself is by no means always vague, and can present quite specific symptoms. "It makes me feel older," is how a middle-aged villager in "Sabay" described viewing the Fantasy Elephant Club across an otherwise familiar strait.

3. PARROTFISH

Frozen parrotfish imported from the Caribbean are for sale in south London fishmongers. The *Scaridae* are specifically reef fish: The beaklike fusing of the front teeth which gives them their popular name has evolved so that they can browse on corals. The parrotfish does indeed damage corals, just as a blackbird damages worms. What counts is the balance of such activities, especially in an ecosystem as delicate as that of a tropical reef. Pulling parrotfish out of reefs in quantities sufficient to supply an export industry cannot but have an unbalancing effect. Through a complex chain of relationships the reef will inevitably suffer from their absence, as it certainly will have from the methods used to catch them.

This is doubly exceptionable, since the members of this family make indifferent eating unless at the moment they are caught. Then the otherwise uninteresting meat is delicious raw, marinaded for fifteen minutes in lime juice, oil, tiny native onions and black pepper. Frozen into woody curves in a bin in a Peckham fishmonger's a parrotfish merely pretends to the exotic, its blunted colors ghosts of what they were. Who knows whether their corpses are there to satisfy the nostalgia of an immigrant population or the local consumer's fickle desire for novelty? Either way, a reef off Barbados or Jamaica is now undergoing a change which will most likely be permanent.

Axel Heyst's "magic circle" would just have enclosed Manila, Saigon, Singapore, Surabaya and Ternate, as well as the Sulu and Riaw-Lingga archipelagoes. (The Mergui Archipelago lies somewhat outside.)

The ocean's emptiness appals the swimmer, but only because it can supply nothing for his own survival. He cannot entertain flabby polemic about dolphins. His is the mind of a man lost in the sea. Yet even as he struggles to save himself he is hollowed out by despair. What is it that he is saving? The thought corrodes his every intention. In this wide salt world which he treads he is nothing, has nothing but a face mask and a pair of trunks. Until one loses everything it is never clear what it was one had. Now, in a bleak inner glimpse, he finds he has dissolved. The landscape of his own past, his private history, seems to have vanished, leaving only a sense of attrition. As he glances down through the water his body dwindles whitely like a distant peg and sheds a small discolored puff of urine which briefly unravels itself in thready convections like those of lime juice being diluted. Nothing but ocean. His entire body is dissolving, too. He only ever existed as three tenths and that fraction is melting into water.

However, this 30 percent contains an animal which does not want to die. A passive animal, maybe, but still perversely convinced that help will turn up as if by more than mere chance. Sooner or later someone surely has to pass within hailing distance of the psychic beacon he must have become, broadcasting his distress signal on all frequencies. He squints at the sun. Now that he no longer wants it to be stuck vertically at noon, it seems reluctant to move at all. Night with its hope of fishermen is still many hours away.

The swimmer tells himself he need not bank only on them. He has been overlooking all the other sorts of boat which continually cross these waters. Besides tattered interisland launches there are all the craft which used to fetch up on "Tiwarik": friendly gunrunners, wanderers from the south with their faces wrapped against the sun, poverty-stricken vagabonds neither peaceable nor violent but chance-takers of more or less competence. Any of them might spot him from miles away with a vulture's quick eye for a weakening beast. He tries to imagine into being a huge arch of cloud letters in the sky: REWARD!, and underneath a gigantic arrow pointing straight down whose tip balances on his sunburned head. It is a message aimed impartially at any of the seagoing mavericks who still inhabit this last corner of the ocean.

So hard does he will it that he soon thinks he hears, above the infuriatingly loud slop of wavelets, the faintest putter of an engine.

VII

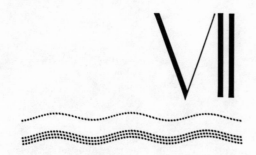

PIRATES AND NOMADS

1. PIRATES AND NOMADS

It was a morning of flat calm in the Sulu Archipelago toward the end of November 1990. The sun in its early angle was both gentle and powerful, forcing a luminous violet into the water. The islands dotting the sea to the horizon stood clear and delicate, as though the same gentle power had hatched them out overnight and left them in their freshness to harden off.

A small boat with bamboo outriggers was heading back toward Subuan. In it were three young Bajau women and a double row of gray plastic jerrycans of water. Perched on these amidships was a seventeen-year-old boy, the brother of one of the girls, who understood the ten-horsepower Briggs & Stratton engine. He was controlling the speed by means of a length of nylon fishing line tied to the throttle. This ran over a rubber disc cut from the sole of a sandal and nailed to a thwart. The boy had wrapped the end of the line around his finger and was concentrating on holding the engine's note at the same pitch. It was a nearly new motor whose exhaust was rigged in local fashion from an iron elbow bolted to the exhaust port with a foot of ordinary water pipe screwed into it and jutting up at an angle. From this unsilenced tube blared the fumes and racket familiar to those on board as a soothing part of a ritual which, before the coming of the engine, had been a much more laborious business involving paddles and a sail made from rice sacks.

The Bajau were homeward bound from the island of Siasi, whose town was the nearest reliable source of fresh water. There was also an acute shortage there, to be sure, but at least water could be bought at a seasonal average of about twenty cents per jerrycan. At first light they had reached Siasi's jetty with a load of fish, which the women's husband had caught the previous day and had kept alive overnight in rattan containers suspended in the tide beneath the floors of their huts. All the Bajau's transactions had been carried out by the one woman among them who knew how to use money. None of the four was literate, but she did understand coins. As long as everything was done in coins it was all right. She would have nothing to do with paper money. Not only could she never be sure of the denominations; it was all too evidently flimsy stuff. Drop it in the sea and it would be reduced to pulp and where were you then? For all her knowledge, though, she and her companions were happiest with barter. Barter, now, that was the proper way of doing things; taking into account all sorts of subtle variables like quality, whether there was a glut or a shortage that day, how much of a hurry you were in. Thus a basket of medium blue crabs might fetch two baskets of good cassava, while a "half-boat" shark half the length of the *bangka* was worth quite an assortment of nylon line, fish hooks and petrol, plus (especially if the Chinese shopkeeper was buying jaws as well as fins) two pairs of children's shorts.

As the boat rounded an islet they could see Subuan in the distance with its detached clump of huts standing well offshore on legs anchored in the corals. The sea's surface glassily collected the white clouds which, toward midday, might with any luck heap together and wring some precious rainwater out of themselves. The Bajau stared forward, reading the water. Probably no other people anywhere could gaze with such knowledge of what was happening below its surface. The sudden swirl of a tail, cat's-paws and dimples of wind, the alignment of the snapped blades of *Thalassia* sea grass whose floating debris threaded the archipelago like oil slicks, even passing smells: All carried information or stood for omens.

No doubt the hammering exhaust and the mesmeric calm prevented the women and the boy from noticing the bigger and faster craft with double outriggers slipping out from behind the islet as they passed. It always would have been too late for them in any case. They would certainly have seen it as it drew level, matching their speed a few yards away. The boy at the engine would have turned, half stood, shocked but not really bewildered by the sight of a man steadying an M-16 across the bigger craft's low roof. And thus, probably without shaping a clear thought, he took leave forever of his sister and companions, of the glittering morning ocean, of his seventeen years. The shots carried away much of his head. His body must have fallen across the scorching engine, for when it washed ashore it was still possible to make out extensive blistering of chest and stomach. His dead finger slack, the engine would have slowed abruptly to tick-over. The little boat lost way at once, rocking gently out in the middle of the sunlit strait, steam rising as the boy's blood hissed and bubbled on the cylinder head.

It is unlikely the three women remained sitting mutely while men swarmed aboard across the outriggers. Being Bajau, their instincts would surely have been to put themselves into the hands of the sea and its spirits. But sooner or later, no matter how ably they swam and dived, they would have been wrenched from the water and tied up on the larger craft. A man aboard their *bangka* unwound the nylon from the boy's finger, hauled his sizzling body off the engine and threw it overboard. Then the faster boat took the other in tow and, veering away from Subuan, probably headed southwest to Tapaano or even Sugbai, which lay on the horizon a scant forty minutes away.

Once there the pirates would doubtless have been joined in camp by their outlaw companions, taking it in turns to rape and beat the women unconscious. At length someone would have set off with all three victims' sarongs (that all-purpose garment which serves variously as skirt, trousers and turban) and arranged to have them delivered to the respective husbands, together with ransom demands of just under £1,000 ($1,755) per head.

Since everyone knew the Bajau were nearly all subsistence fishermen and the poorest tribe between Mindanao and Indonesia, it is not clear how seriously these ransom demands were intended. Maybe the pirates thought that if a poor Bajau could put an engine in his *bangka* he had a secret source of funds which might be further tapped. In point of fact the motor had been installed by a Chinese, a merchant keen to boost his trade in marine products which he periodically shipped to Manila. In any event he was not a man to throw good cash after a lost engine. After all, no one seriously thought a Bajau was worth . . . well, how could one put a figure on the lives of illiterate fishermen, gypsies who mostly lived outside any sensible economy? Still less on their wives.

And so, after a week's captivity, during which time no money arrived and their treatment no doubt reflected the pirates' increasing anger, the three Bajau women were put to death in a way which can only be imagined. All that is known is that the body of one was chopped up into quite small pieces—*diced,* it would be called in a recipe—and piled in the bilges of their little *bangka* over the four bolts on which the Briggs & Stratton engine had been mounted. Then it was set adrift. The boat with its heap of meat wandered with the current for a while before being found, recognized, and read as just one more awful warning in a region used to awful warnings and worse deeds.

This small atrocity was still mentioned now and then by the island folk three months later, although by that time it had been overtaken by news of more recent acts of piracy. Such things were commonplace. The least advantage in material goods or business put anyone firmly in an extortioner's sights. Uncharacteristically, these pirates had forgone a *bangka,* but they had gained a nearly new engine, some full containers of drinking water and a week's fun. The world spun on.

As for the Bajau themselves, they would not have forgotten. At least they were able to take the boy's burned and headless body to their cemetery island and lay it with semipagan rites beneath a carefully painted wooden board, surrounded by gay little flags. Then, as is their custom, they probably raced each other back down the beach and into the water to wash, spurred on by the half-serious belief that "the last one in is the next to die." Perhaps they instinctively felt the sea safer than the land. Maybe some distant tribal memory warned them that, like turtles, they were doomed to transact certain necessary rituals on dry land before they could once again return to the comparative safety of the ocean.

~

Such an anonymous event—which was never reported in any newspaper, nor formally to any military or civil authority—contained within it many of the well-worn coordinates of the Bajau's fate, not merely of three women and a teenage boy but of the whole scattered tribe. Come to that, it was characteristic of an entire region. Everything that had taken place was immemorial within its setting: the fetching of water, the bartering of goods, the being victim, the being pirate. So also were the lumps of land rising from the seabed haphazard of all demarcations, the shoals and atolls, sandbars and islets drugged with sun as the archipelago (a word whose beautiful syllables stretch themselves to the mind's horizon) sprawled in its great tropic swoon while seething with violence. Many of Conrad's best stories were set in this area. From southern Mindanao, from Sultan Kudarat and Zamboanga and Palawan down to Java in the south, from Sumatra in the west to the Moluccas in the east, certain things have changed little to this day. Some names are different. Celebes has become Sulawesi, Makassar is Ujung Pandang and Batavia is Jakarta. But the rest—Ternate, Surabaya, Kuching, Samarang, Timor—still exist and echo with the pungency that thrilled my adolescence until like Axel Heyst himself I could believe my life enchanted by a magic circle "with a radius of eight hundred miles drawn round a point in North Borneo."[1]

The way of living is still much the same for many in the Sulu Archipelago. They have not stopped diving to eardrum-splitting depths for pearls and *tripang* (sea cucumbers); the blue highways of water are patrolled as ever by sharks and crisscrossed by inter-island traffic of every kind. The pirates, smugglers and common cutthroats are very much in evidence. Even vestiges of the old sultanates remain, although the turbaned despot living in his fiefdom—a pocket trading empire defended by mangroves, a treacherous estuary and riverbank spies—tends nowadays to be a pretender who manages a grocery store in Jolo and writes long letters by candlelight to the United Nations, beseeching it to recognize him as rightful heir. Everywhere the kris has given way to the Armalite, while the great white sails of clippers and schooners have been supplanted by Isuzu marine diesels. Otherwise, Conrad might recognize these waters as having retained their archipelagic essence: seductive,

---

[1] Joseph Conrad, *Victory* (1915).

dangerous, possessing above and below their surface a treacherous quality which leaves nobody untouched.

He would certainly have been familiar with the Bajau, although in his day they were more exclusively boat-dwelling than they are now. Anthropology has still not decided where they originally came from, or why. The earliest European visitors, beginning with the Portuguese and Magellan in the first quarter of the sixteenth century, reported small groups of boat-dwelling nomads throughout the archipelagos of Southeast Asia. It is largely a matter of guesswork as to how long they had been there. Opinion is divided, too, on whether they once all came from the same tribal stock—whether, for example, the sea nomads of the Mergui Archipelago off the western coast of Burma share a common ancestry with those of the Riau-Lingga Archipelago in Indonesia and the Bajau of Sulu and North Borneo. Whatever their origin, it is always presumed that at some point far back in time they were all land-dwellers who for some reason decided to go and live as far as possible afloat. What is now to be seen is the final phase of this ancient way of life as the Bajau try to take up residence on land again. The difficulty they are experiencing in doing so is a measure of the thoroughness of their social adaptation to living on the sea.

One of the theories as to why their ancestors left the land in the first place is that they were literally driven off it by stronger tribes. If so, it is ironic that similar persecution is now driving the Bajau off the sea. Three tribes predominate in Sulu: the Tausug, the Samal and the Bajau. For several reasons, principally linguistic, modern opinion tends to lump these last two together. The Bajau language is a dialect of Samal (properly called Sinama) and besides, they consider themselves to be Samals of a kind. "Of a kind" is a reasonable qualification, since the Samal of the southern archipelago are a heterogeneous bunch and can vary in dialect, attitudes and social customs from island to island. At one extreme they include illiterate, pagan, boat-dwelling Bajau and at the other sophisticated traders and teachers who have made the hajj to Mecca.

Ranged against all of them are the Tausug. The Tausug have always been politically and otherwise dominant in the archipelago, with a reputation for pride, hot temper and general disdain for lesser folk such as Samal. As for the Bajau, they often refer to them as *luwaan. Luwa* is the Tausug verb for "to spit out"; "outcast" would probably be the nearest

English equivalent. No doubt a good deal of Tausug bigotry toward the Bajau is caused by—or at least explained as—a matter of religion. The Tausug have a generally high opinion of their own version of Islam, rather less of the Samal version, and can express serious doubts about whether the Bajau brand even counts as Islam at all. It would be impossible for a people living in a predominantly Islamic area not to have assimilated a great deal of Islamic culture, and the Bajau have done so, many of them being devout Muslims. But since they are traditionally a boat people, they are perforce a people without architecture. Hence their mosques—for all the world dilapidated huts on stilts standing in the waves—do not fulfill the Tausug idea of what a mosque should look like. "How can a people without decent mosques qualify as Muslims?" I was asked by some exceedingly hospitable Tausug on Siasi. What I should have countered with, had I not lacked the requisite nerve and discourtesy, was "How can a people drink as much as you do and qualify as Muslims?" for drunkenness in this famously Muslim region is widely admitted even by Imams to be one of the principal reasons for the extraordinary level of casual violence. It was, of course, Tausug pirates who killed the three Bajau women and the boy. Although it is quite certain their Bajau relatives could identify the men involved, they would have been much too frightened to report them even had there been any form of law in the area worth reporting them to. Seen from far enough away, of course, the Tausug and the Samal and the Bajau appear to have far more in common than not; seem practically indistinguishable, in fact, since they are all people whose lives are overwhelmingly mediated by the sea.

Left to himself, the Bajau's is a peaceful, shy, somewhat nomadic way of life which is as highly specialized as that of the Eskimos. It would be a mistake to romanticize it, though. Bajau living has always been extremely hard, often dangerous, lonely and beset with disease. The traditional family unit was a married couple and their children in a low-roofed boat. Depending on season, preferred fishing grounds, family events (mainly marriages and deaths) and sheer whim, the boats either formed parts of flotillas, floating villages, or went their own separate ways for months on end. They tended to put ashore only for drinking water, firewood and cassava; otherwise it was a life at sea in all weathers. It was also a life largely spent crouching: squatting to cook, squatting to eat, squatting to fish. To this day older Bajau can still be seen who walk on land with a gait as characteristic and graceless as a duck's, their lower limbs slightly atrophied from a lifetime's hunkering down.

The conditions in these boats, where they still exist today, can become squalid, to say the least. It is not merely that if there are infants and toddlers aboard the bilges soon give the craft the atmosphere of a floating urinal. All sorts of vegetable ends, cooking scraps and fish guts fester there as well, brewing up in the tropical climate so that the liquid, when glimpsed between plank and thwart, can be seen fizzing. This is baled out at anchor in shallows, but on the high seas is often allowed to build up, and for a perfectly good reason. Since Bajau men spend a good deal of time overboard cutting seaweed and diving for *tripang* and pearl oysters, the last thing they want to do is lay a trail of offal and attract sharks. This would be a determining factor for any sea-dweller in the tropics and, indeed, is noted in a curious account by an Englishman of trying to run a timber business in the Mergui Archipelago in the 1920s.[2] The sea gypsies he knew there, and whom he tried to introduce to the idea of paid labor on land, called themselves "Mawken," which he translates as "seadrowned." The boats and habits of the Seadrowned People of the Mergui Archipelago seem to have been not much different from those of the Bajau, which is hardly surprising given the similarity of the conditions. The Mawken, too, gave an account of their origins which told of persecution by Burmese hill tribes from the north and Malay pirates from the south squeezing them off the land and into the sea. They also offered this as an explanation of why they had no interest or skill in cultivation, merely picking up fruit or an occasional wild pig when passing an island. According to Ainsworth the Mawken seemed to exist on seafood, rice and opium. The Sulu nomads mostly eat cassava as their staple, only the wealthier tribes and classes relying on rice. I never saw opium, but marijuana is widely smoked.

Malnutrition is a common consequence of this way of life, partly because the diet is unvarying but also because the choicest fish are mostly reserved for sale, while the fisherman himself subsists on scraps, shells and the coarser varieties. Combined with the lack of sanitation this leads to a high incidence among the Bajau of tuberculosis, malaria, infant

[2]Leopold Ainsworth, *A Merchant Venturer Among the Sea Gypsies* (1930).

diarrhea and infestations, to all of which their resistance is slight. This seems ironic, even contradictory, in view of the extraordinary physical fitness and imperviousness to cold needed by men free-diving to up to 100 meters for pearl oysters. Even greater depths are claimed; and although so far as I know nobody has ever bothered to measure them exactly, such dives have to be placed against the current world record of 105 meters,[3] especially since these are working dives during which shells, lobsters or sea cucumbers are gathered. Before diving the men often swig a mixture of canned milk beaten up with little "native" eggs and lemon juice, and eat bananas. They say this makes them resistant to the cold.[4]

Young Bajau, particularly children and adolescents, often have the bleached, tawny hair of people who spend their lives in and out of the sea. Some even approach a dusky blondness which, amid uniformly black Asian hair, is very striking indeed and a sure guide to that person's way of life and social status. No doubt it is Western culture, and specifically cinematic images of Californian or Australian beach culture, which equates this with the very pinnacle of healthy living (or did before skin cancer was mentioned). In Southeast Asia, though, it speaks of a life of poverty and often malnutrition lived beyond the least shadow of a classroom, with intestinal worms and scabies as constant companions.

Even in the sixties, when anthropologists like H. Arlo Nimmo were undertaking classic studies of the Bajau, their landward trend was obvious. The entirely floating life on houseboats was being relinquished for a tentative one in stilt huts which represented the placing of a wary toe on the very edge of land. Like their Samal cousins, certain Bajau always did build communities of huts on stilts in shallow offshore areas, each hut joined to the next by sagging walkways. The difference between them was that while Samal villages always had a gangway leading ashore, Bajau villages did not. There were other signs, too, of a culture becoming less isolate.

[3]*Guinness Book of Records* (1991). This record was set in 1983.

[4]For this and many details of Bajau life I am indebted to Dr. Saladin S. Teo, a native of Laminusa Island, Siasi. In addition to being a superintendent of schools in Sulu, he has made a particularly sympathetic study of the Bajau over many years. His book *The Lifestyle of the Badajaos* (1989) is a useful addition to the literature; but even more valuable to me was his friendly and courteous company on visits to Siasi and Laminusa.

'Paganism' and the boat-dwelling habit have always been identifying marks of the Sulu Bajau. With the full acceptance of Islam and the abandonment of the nomadic boat-life, these sea folk will cease to exist as a 'pagan' outcast people, and become amalgamated into the general Muslim Samal population of Sulu. Probably within another decade full-time boat-dwellers will disappear completely from the Sulu waters.[5]

This was what Nimmo wrote a quarter of a century ago, and since then the drift to land has become almost total, except for a few isolated cases. Yet his prediction of "amalgamation" has certainly not come about, if this means the adoption of land-based social habits and values. To this day the whole problem of how the Bajau can be integrated remains unsolved. They are mostly unconvinced by the idea of education, so are often unwilling to send their children to school. Nor do they seem keen to learn new skills. And as for taking part in any social or political activity, it has proved almost impossible to interest them. They suffer, in short, from an admirable lack of ambition. Their relationship with the sea is so strong they give the impression of being only flimsily attached to land, and might leave again tomorrow if conditions became any worse. Maybe the sea itself is by the way; perhaps what they have in their blood is a nomadic indifference to roots. This might explain their amiable remoteness, their strange innocence. Since they have never owned property ashore, they have always remained free of contaminating land squabbles, battles with landlords and developers, crippling rents and tribal annexations. At the first sign of trouble an entire settlement of Bajau may be discovered to have left overnight, in silence, their abandoned hut creaking slightly in the tide and their low craft already invisible over the horizon.

What has done most in the last twenty years to change the Bajau's way of life is violence. No anthropologist writing in the sixties foresaw that persecution would increasingly drive them ashore, and that the shore with its press and clutter of people, its social crosscurrents and complexities would prove a very mixed blessing. Nobody guessed they might have to inhabit a strange no-man's-land, an intertidal zone. But then, nobody realized to what an extent the Sulu Archipelago would become a battleground. In 1974 Ferdinand Marcos sent in the armed forces of the

[5]H. Arlo Nimmo, "Reflections on Bajau History," *Philippine Studies* 16, no. I (1968).

Philippines against the Moro National Liberation Front. In the fighting of early February that year, most of the town of Jolo was destroyed and its population forced to take to the hills. Henceforth, the best that reigned in Sulu was armed stalemate broken by violent guerrilla and military engagements, until today's state of undeclared anarchy was reached. The great influx of weapons into the area, together with financial support for the MNLF from abroad as well as the money brought by lucrative smuggling and trade links with Malaysia, meant that the dominant tribe became more dominant still.

In the last ten years anyone has been able to acquire an M-16. If all else fails, one can easily bribe one's way into the Army and acquire the weapon that way. M-16s are constantly "lost" as soon as they are issued, and often the new recruit only waits until the weapon is in his hands before defecting. Armed with an M-16 it is simple to steal a boat. Anybody with an M-16 and a *bangka* can go straight into business on his own account as a pirate. It is a vilely dangerous living, to be sure, and the sharks must have grown very fat in the straits between the islands, but it is a way of life sanctioned by tradition and facilitated by the times.

The result for the unarmed, peaceable Bajau has been disaster. Whereas once they could fish at night using hurricane lamps and Coleman lanterns they now dare not, for fear of attracting pirates. A further disadvantage is that the pirate craft frequently have engines powerful enough to outrun the Philippine Navy and coastguard patrols and whereas once the Bajau might have used superior seamanship to avoid trouble they are now helpless. So if anything has reconciled some of them to land and its unfamiliar ways it is the need to defend themselves. It is not Islam, nor free education by the Oblate Fathers, nor offers of health care, nor any amount of blandishments and promises which have changed the Bajau's horizon in Sulu. It is violence, and the necessity of earning enough to buy an M-16 and an engine in order to counter it.

~

Presumably, nomadism—whether of Bedouin or Eskimos or sea gypsies—is everywhere in decline. In order to survive, nomads need large tracts of unoccupied territory where there is no serious competition for their source of food, and such areas must be growing fewer. Besides, modern governments increasingly dislike "floating populations" who seem ignorant of their control and who drift uncaringly across their borders and frontiers as if they did not exist. All centralization is a threat to the periphery, and minor tribes which fall outside even the periphery tend to become fair game. It is a short step from being a minority to becoming marginal and then officially outlawed.

Even under average conditions nomadic life is harsh, while a single stroke of ill fortune such as drought, epidemic, civil war, an oil spill or volcanic eruption can bring a people to the edge of extinction. To sentimentalize nomads is not a patronage they need. With an Armalite at last in their hands and deep memories of catalogs of injustice they do not necessarily behave better than anyone else. What they retain, though, is priceless: a genuine remnant of the knowledge which has served the various species of *Homo* throughout his history. This knowledge is already lost to industrialized man and within the next century will be lost to the human race for good. It is a particular way of being in a landscape, of coexisting with ocean and land which takes account of minutiae we no longer know how to observe and maybe now cannot see at all.

There is a link between nomads and pirates and even smugglers. It has partly to do with living in a world beyond boundaries, but also with a detailed knowledge of that world which goes beyond mere geography. Pirates are simply seagoing versions of highwaymen or brigands; each calculates that his knowledge of the locality will be superior to that of any forces officialdom sends out to capture him. But in between moments of intense danger and excitement must be stretches of considerable solitude, and some sea pirates must themselves have a near-nomadic existence. After all, piracy need only be a sideline. At its lowest level, such as that which has all but driven the Bajau to land, it is a matter of rat-poor fishermen preying on other rat-poor fishermen for the simplest things, like a day's catch or a dugout boat. I am sure that half the wanderers who landed on "Tiwarik" when I was there were neither particularly fishermen nor pirates nor smugglers but all three as occasion demanded. They might be best described as opportunistic nomads, and what characterized them all was that they were highly self-sufficient. It was not a luxurious life they led, but they were utterly at home in it. First and last, they were born boat people. All had that adhesive agility common to those who grow up barefoot on very small craft. Most had the tawny streaks in their hair, the bleached expression and frown lines of those who have squinted constantly at glaring horizons. All were skilled with dynamite, hook and line or woven fish traps. None was scared of man or beast, but they were truly frightened of *mumu,* sea spirits and omens.

In its immense navigational complexity and its lavish range of hiding places, a tropical archipelago is ideal pirate territory, and piracy has been established for centuries in insular Southeast Asia. Some pirates achieved fame and most Filipinos know the story of Lim Ah Hong, the Chinese pirate who in 1574 even raided Manila itself and came close to unseating the fledgling Spanish colonial administration. His name lives on, less for nationalistic reasons than because of a vast treasure he allegedly hid and which has been looked for ever since. (A treasure is, of course, any proper pirate's true legacy.) Down in Sulu, in Borneo and the East Indies, piracy always flourished well. This was partly because it was Muslim territory, with a complex assortment of fiefdoms and sultanates never brought under full control by any colonial power. When various Sulu potentates made alliances with their counterparts in Mindanao the entire Philippine archipelago became prey to Islamic pirate junks. The more regular the colonists' shipping and trading became, the better the pickings, until by the nineteenth century piracy had reached epidemic proportions. "From Mindanao to Sumatra, countless White travelers recorded their fears of, and warnings about, the savage marauders of the archipelago who thrived on massacre, violation and rapine."[6] In 1830 Stamford Raffles himself had found "no vessel safe, no flag respected."

Today's predominantly Tausug descendants of those pirates who infested Sulu are merely carrying on a long-ingrained tradition. Naturally, piracy can hardly thrive without victims, and in default of galleons carrying Spanish gold from Mexico there are interisland launches carrying people with wallets and cargoes of goods for Chinese traders. There are also the boats which run regularly between Jolo and Labuan Island in Malaysia, taking advantage of recent barter trade agreements under which copra and handicrafts are swapped for electronic goods, textiles and canned food. It may sound small-time, but each round trip can be worth up to £100,000 ($175,500) and certainly those concerned take it seriously enough. At the very end of January 1991, pirates killed twelve Sulu barter traders in a single concerted raid. As for smuggling, there is a brisk trade out of Sulu in marijuana, which also goes to Malaysia. This

[6]V. R. Savage, *Western Impressions of Nature and Landscape in SE Asia* (1984).

seemed unlikely enough, given that country's famously draconian penal-ties for drug dealing; but as I was succinctly told, "Malaysia's a big place." In return, "blue seal" American cigarettes are smuggled back and are on open sale throughout Sulu and Zamboanga.

These are not romantic businesses to be engaged in, and certainly not to fall foul of. The treatment of the Vietnamese "boat people," the refugees who fled Vietnam after 1975, was a case in point, and those victims who lived to testify to pirate attacks—often by well-equipped Thais—gave horrendous accounts. The earlier waves of refugees were largely Chinese middle classes from the Cholon district of Ho Chi Minh City, lately Saigon, and were often wealthy. They brought what concen-trated valuables they could with them, usually gold, hidden about the overcrowded boats with great ingenuity, sometimes suspended by brass wire beneath the keel. Maybe the earlier pirates were satisfied by good hauls, but as time passed and the Chinese were replaced by ordinary Vietnamese political refugees their savageries began to be mentioned in the world's press. All in all, it was a far cry from the behavior of the Bangladeshi pirates who in November 1989 were reported as singing their victims a little choral medley before asking them to turn over their valuables.[7]

Instances of horrible and immensely daring opportunism abound among the archipelagos of Southeast Asia and are frequently evidence of a nomadic understanding of the sea. Less than a year after the *Doña Paz* disaster another overcrowded Sulpicio Lines vessel, the *Doña Marilyn*, sailed for Manila out of Cebu despite the coastguard's warnings of the imminent arrival of typhoon "Unsang." On the night of 26 October 1988 the *Doña Marilyn* sank while trying to shelter in the lee of Guiguitang and Manok-Manuk islands off the north coast of Leyte. On this occasion, at least, there was land nearby. The seas were very heavy and many survivors who managed to swim in the right direction were pounded against the jagged offshore reefs and died there. And yet while dozens of brave Manok-Manuk islanders formed a human chain far out into the surf to pull exhausted swimmers in, other villagers who had heard the ship's radioed distress signals and had seen her lights had long since launched their flimsy *bangkas* and were far out in the storm, hauling survivors aboard, stripping them of any valuables, and throwing

[7]BBC World Service, *Meridian*, 11 November 1989.

them back in. Had they been rescuers, their courage could scarcely have been overpraised. Yet as plunderers their bravery was actually no less. They displayed a true piratical streak that night, amoral and enterprising.[8]

To make a living from smuggling, as from piracy, one needs to know the territory with a precise and local eye. This must be true whether on land or sea. In regions where particular trade routes run or particular economies have grown up, smuggling activities will develop their own skills, lore and traditions. The bootleggers of West Virginia who ran illicit corn liquor through the Allegheny Mountains developed driving and engineering skills for outrunning the law which in turn nourished the entire sport of American stock-car racing. Mountain boys drove as soon as their feet could reach the pedals, and apart from learning a repertoire of tricks (such as the "bootlegger's turn") they also acquired great sensitivity to details of road surface, weather conditions, and a car's balance and handling depending on how full the hidden tank of liquor was. The archipelagic people of Southeast Asia have analogous skills, but they have others as well which make land-based versions look coarse and two-dimensional. Above all they are prodigious navigators.

The Bajau's ability to go back to a particular patch of ocean without reference to land seems uncanny. Stories are told of their being able to sail unerringly to a single lobster pot on an overcast night out of sight of land. I have never seen this, but certainly such things are habitually said about other sorts of nomads, whether Aborigines in Australia or Kababish camel herders on the fringes of northwestern Sudan. They are peculiar to any people whose entire living depends on a knowledge of

---

[8]To forestall any moral posturing it is salutary to go back half a century to when London was behaving gallantly during the Blitz. On the night of 8 March 1941, a Saturday, two bombs fell on the crowded Café de Paris in Coventry Street which was jammed with couples dancing to Ken "Snake Hips" Johnson's Caribbean Band. Since the place was deep beneath the Rialto cinema it was thought of as safe, but the two bombs went right through and into the nightclub, where one exploded, killing thirty-three outright and wounding sixty, while the other merely split open and sprayed bright yellow picric acid over everyone. In the semidarkness, choking fumes, dust and general carnage, two men posing as members of a rescue squad went round calmly removing rings from the fingers of the dead and unconscious and turning out handbags. In fact, there were organized gangs who had an elaborate telephone network keeping them up to date on where bombs had fallen and which places might present the best opportunities for looting.

their natural surroundings and who are themselves largely bound into the ecology of the area. The Bajau's knowledge of the sea comes as much from living in it as off it and extends to its every aspect.

~

Anthropology has confirmed what was self-evident long before Thor Heyerdahl and his *Kon-Tiki* venture in the late 1940s. That is, that sea nomads have always been serious navigators. Fifteen hundred years ago the Polynesians were sailing around the Pacific in big catamarans using the stars, frigate birds, sea conditions, smells and their own stick maps to tell them where they were. (It would be interesting to know if these stick maps, whose intersections marked islands, shoals and currents with considerable accuracy, also marked *imaginary* islands which, over the centuries, gradually disappeared.) Recent research concludes that *Homo*, like many other species, does have a built-in sense of direction, no matter how atrophied it may have become from disuse.[9] Apart from navigation, though, a sea gypsy's knowledge of the ocean is scientific in its detail, yet his is very far from being a scientist's gaze. For one thing, it tends to be holistic to a degree, whereas the impression given by most of the geophysicists aboard the *Farnella* was one of extreme specialization.

The question finally asks itself: What order of knowledge do we stand to lose if and when such people as sea nomads finally abandon their way of life, and does it matter? Perhaps one can say with more than mere intuition that certain innate skills and faculties do atrophy through not being used, that an increasing reliance on electronics to mediate our apprehension of the world does lead to the loss of certain sensitivities, and that to lose any sensitivity or awareness is limiting and unwise. Again, extreme examples are sometimes advanced in favor of retaining "old methods." In the case of navigation, for instance, it might be said that with increasing reliance on satellite-based positioning and guidance systems, the old skills of stellar naviagation may no longer be taught even as a "manual backup," and will in time be lost altogether. "What happens," the argument runs, "if something puts all electronic naviga-

---

[9] The evidence comes from laboratory experiments during which subjects' ability to distinguish North gradually improved, which "suggests that orientation in humans is a latent sense, which in some people can be recalled very successfully after multiple challenge" (Mary Campion, *The Journal of Navigation* 44, no. I (January 1991).

tional systems out of commission at once? Suppose there is a massive solar flare whose radiation disrupts the Global Positioning System satellites?[10] Or one of those sudden reverses of Earth's magnetic polarity which would make it necessary to recalibrate all compasses? What then?"

Of course, this is not quite the point, though there is a poignancy in watching the old and new technologies meet. In the early seventies I found myself flying from Recife in Brazil to London's Gatwick Airport in a VC 10 of British Caledonian. The aircraft was virtually empty. I was one of eleven passengers, and after the others had settled in for the night (they were mostly elderly) I was invited to spend as long as I liked in the cockpit. Such innocent, preterrorist days they were; casual in the economic sense, too, which is no doubt why the airline no longer exists. In the middle of the night the navigator, who had been getting radio fixes from Dakar and Cape Verde, stood up and opened a tiny hatch in the cockpit roof which he called the "smoke ventilation hole." this exposed a plexiglass bubble through which he shot the stars with a sextant. Now, twenty years later, there are no more navigators in airline service, the last having flown on VC 10s and Boeing 707s. The crew on the flight deck of a Boeing 747 consists only of the captain and the first officer. Neither has been trained to navigate by the stars. Nor has the cabin crew. If an aircraft is forced to ditch and its passengers and crew manage to haul themselves into the rubber dinghies they will not, unlike Captain Bligh and his fellow officers, be able to make a dogged landfall weeks later nor even, like downed World War II aircrews, know which direction to paddle in. All they can do is sit impotently bobbing up and down, waiting for rescue.

The point is not only what will happen if and when stellar navigation becomes a lost art, but who apart from astronomers will remain attentive to the heavens? And who apart from scientists will remain attentive to the sea? Even when it happens before our eyes it is hard enough to accept that species become extinct, that they always have and always will, since without extinction there is no evolution. But the idea of bodies of *knowledge* becoming extinct seems quite as shocking, and it is difficult

---

[10]Such flares are not uncommon. A few years ago the electromagnetic energy of a solar storm induced currents in landlines which caused widespread power failures in Canada, blacking out entire cities. Recently, a similar surge of solar radiation was enough to slow several GPS satellites, altering their orbits and hence the accuracy of their information.

to see how it can be avoided when they are so inseparable a part of a rare and specialized way of life. It is too late now to save many a tribe—of Amazonian Indians, for example—who might have spared us years of suffering and expensive research had they been consulted in time about the medicinal properties of the plants they knew best. (This, of course, is the utilitarian approach to conservation.) Maybe, after all, bodies of knowledge peculiar to a tribe should, like species, be allowed to become extinct once circumstances have changed and they can no longer adapt themselves.

Apart from rebelling instinctively against it, it is not an easy argument to counter. If in fifty years' time most Bajau are stockbrokers, what will the sea be to them except somewhere for family outings and expensive water sports? Of what use to future generations their present intricate understanding of the ocean? If there is a scientific rather than a sentimental answer it might be one analogous to that which sees the paramount importance of maintaining the diversity of species, of the gene pool. The more the world becomes politically, economically and culturally centralized, the more homogenized its ways of living, so the dangers of sameness become apparent. To take a notorious example, the EC regulations restricting the varieties of seeds permitted for sale within the community have for years been viewed as potentially disastrous by botanists and agronomists. A real threat is concealed in the preferencing of a handful of crop varieties chosen only according to marketplace (mainly visual) criteria. Once the genetic bank is depleted the chances of calamity caused by a single unexpected virus or pest become much greater. When in the nineteenth century the Irish potato crop was lost, creating mass famine and emigration to the New World, the potatoes were almost entirely of a single strain, uniformly susceptible to blight. In future, no amount of genetic juggling or selective pesticides will be as effective as growing the widest possible variety of fruit and vegetables, keeping unfashionable strains alive even if the immediate benefits are not obvious.

A consumer-based cultural uniformity is still some way off, but is already advanced enough for certain grim futures to be imagined. At the same time, utterly various ways of experiencing the Planet still do survive, though tenuously and in scattered fashion. The Bajau looks up, and the sun crossing the sky tells him any number of things, among them his place, his time, and how the sea creatures on which his living depends will be behaving. In another world entirely, one spanned with satellites and a global money market, the sun is just a noun, a hot and dazzling object rising with the Nikkei and setting with the Dow-Jones.

2. POEMS TO TECLA

From time to time one notices around offshore corals, and sometimes even far out to sea, small insects skating on the water's surface. Usually they are to be seen in flotillas which mill frantically at the advance of a hand, but lone specimens can be found. They are water striders of the genus *Halobates* ("salt treader"), the only insects out of roughly a million described species to inhabit the open sea. There are thirty-seven coastal and five oceanic species of *Halobates,* and relatively little is known about their biology. Unlike most intertidal animals they have colonized the sea from the land, and face a seemingly hostile environment. They cannot fly; they cannot dive; and to watch them skimming the surface one would imagine they would be crushed by an eddy, let alone by a wave.

The more one examines them, lying in the water or uncomfortably draped over a boat's sharp prow, the more one wonders. It is known they eat the debris of smaller insects: midge corpses, and so on. It is known some of them lay their eggs on the blades of sea grasses and coralline algae. The mating behavior of a couple of species has been studied, as well as their way of avoiding predators.[11] But much remains mysterious, not least how a land insect has found enough in its favor on the sea's surface to make its home there. If such a living can be made, why is *Halobates* the only known insect to have discovered it? And, infinitesimal speck on a liquid desert that it is, how does it ever find the tiny fresh corpses on which it feeds? What is its relationship with salt, and how does its physiology handle the salt economy? What happens to it in the torrential downpours which smash the sea's surface into foam and can put a drinkable layer of fresh water on top?

These are the sorts of questions which time and zoology will no doubt answer. In the meantime one is left with constant surprise at how well forms of life adapt to conditions that seem impossibly harsh or daunting. The dislikable neologism "extremophiles" has recently been attached to them, though to describe an animal as a lover of extremes is an obvious anthropomorphism. Varieties of life colonize any place or set of conditions that will support them; they do not rate them in terms of human comfort. It was precisely these sorts of argument which helped the "azoic" theory last as long as it did. Since then, of course, there have been abundant discoveries of life-forms which have adapted to extreme conditions, including bacteria in active volcanoes; hydrothermal vent communities of crabs, worms and fish living next to "black smokers" on the deep seabed with metabolisms attuned to near-boiling temperatures; lichens and mosses in the Antarctic; desert snails and extraordinary plants like *Welwitschia mirabilis,* whose two curling, strap-like leaves winnow so much life out of blazing heat and nightly fogs in the Namib Desert that single plants may live to be over a thousand years old.

"Extremophilic" even seems a not inaccurate way of describing certain members of the genus *Homo.* It is difficult to see a culture like that of the Eskimos or Bedouin as founded on timidity. Where nomads are concerned it is clear there are certain cultures, as there are individuals, to whom wandering is a necessary part of the economy of living, not merely of survival. Just as people may grow up loving the most nondescript homes and surroundings, so cultures develop which are deeply

[11]See papers by W. A. Foster and J. E. Traherne, Department of Zoology, Cambridge.

attached to apparently unpropitious landscapes and conditions. "Attached to" implies both love and dependence, in which it is not possible to distinguish between a living and the place where it is lived. Deserts and oceans, which to an outsider seem to share great similarities in that both appear virtually featureless and both are life-threatening without specialized knowledge, are places which transcend their own conditions to the point where some people consider them spiritual entities. To wring a living—still more a livelihood—out of them requires skill and courage but also love. In this case a word like "extremophilia" seems, if still graceless, quite appropriate. It is beyond understanding why governments and their agents should imagine that cultures which have taken millennia to attune themselves to such ways of life might cheerfully renounce them in as little as a generation. The world is dotted with groups of demoralized tribespeople, drunk in shacks and shanties at the margins of the societies that have disinherited them. Exasperated and not always unsympathetic officials complain about the inertia and fecklessness of Aborigines or American Indians, how pathetically they connive at their own degradation. Similar officials throw up their hands over the Bajau, baffled by their lack of interest in education or modern health care. But why would a Bajau wish to become assimilated or strengthen his ties with land when his entire history is based on the knowledge that in the long run life on land means nothing but trouble? The sea is not something he can turn his back on to order. To assume he can, or would want to, betrays the origins of the social science that expects it. Western, and particularly American, society thinks little of moving hundreds of miles to take up a new job; but to equate nomadism with mobility is a gross mistake.

When presenting *Homo* with an environment for living, the ocean strips away inessentials other than skill, knowledge and affection. It is a life which requires its own intensity and exacts its own discipline. Many people who are not tribal nomads by birth live lives as remote from ordinary society as that of the wandering albatross. This bird, electronically tagged, has been recorded as making single flights equivalent to the distance of Australia to Britain before returning to a particular island.[12] (In view of a life spent mostly on the wing it would be too cozy to call this place "home.") Among their many human counterparts would be

[12] *New Scientist* 1759 (9 March 1991), p. 55.

the seamen who spend three or more years continuously afloat. One sunny afternoon on board *Farnella* I watched crewmen painting the deck around pink-and-white geologists as they sunbathed. Middle-aged men, they swapped a few Hull syllables with each other and painted as carefully as if the ship were a house they had clubbed together and bought with a mortgage. They paid no attention to the scientists, not even to one wearing a bikini and no top. They gave off an aura of austere contentment, as if they were pleased the *Farnella* was going nowhere while doing so along the most precisely navigated paths. In this way they combined seagoing professionalism with perpetual nonarrival. How trim the ship was! At three in the morning one would come across a man humming to himself, kneeling with a scrubbing brush in a toilet in a cloud of bleach vapor. Val did not think of himself as one of that sort, however, since he admitted to being at sea just for the money. He had lost his woman and sold his house and now wanted to rebuild his finances.

"How else would I make three hundred and seventy-five dollars a week?" he wanted to know. "But it's no fun, I can tell you. Be honest, I loathe the sea. Why won't the bugger keep still? It doesn't seem to bother them. They're a funny lot. Some are here because they've got broken hearts. Ah, didn't know *that,* did you? True, though. Some because they've got no other home. Really, they're only happy at sea. Beats me. They dread going back, and that's true. A couple of days and they don't know what to do.

"It's obvious why they're here, isn't it? Course, it's to avoid responsibility. No mortgage, no insurance, no tax, no car, no electricity bills, no gas bills. Free board and lodging. Those deckhands specially. Been at it since they left school, if ever they went, that is. They don't know any other life. Very ignorant. Very narrow minds, if you ask my opinion, though I don't mean that as criticism. The sea's all they know. They're mostly terrible people when they're ashore. They can't handle it, so they just get pissed. You never see them drunk on board, do you? Bit tiddly, maybe, but never falling-down drunk like they get ashore. No, they must really like the life. I've caught one or two of them sometimes, up on deck, just staring at nothing like a normal person'd watch telly. Mesmerized."

All ocean drifters, "salt treaders," lone yachtsmen, mystics, island-hoppers, wanderers and hermits have a degree of impatience with, and ignorance of, the greater world. A further characteristic of nomads, which many of them share, is an absolute vagueness about geography combined with a precise knowledge of orientation. There is a sense in which no beautifully drawn chart can be made to coincide at any point with the inner maps they carry with them. The two do not relate to each other. I have met tribespeople deep in the western Egyptian desert who had no remote idea whether they were in Egypt or Libya. Nor, so far as they knew, did they hold any particular citizenship. The distinctions they made were linguistic and tribal, and the elaborate kinship tables each carried in his head amounted to maps. I would not be surprised to learn that some Eskimos are much the same, and while always knowing exactly where they are may not know if others happen to call that place "Greenland," "Canada" or "Russia."

There is something irresistible about this, since it affirms the ancient homogeneity of land or ocean, a unity of human experience that transcends temporary political boundaries. As for the seas, those vast tracts, seven tenths of the Planet's surface, of course they are mysterious and haunting. Time and again they draw people back to them: mad mariners, adventurers, solitaries, very often misfits on land who are transformed once they are afloat. Their keels cut tiny scratches on the face of an abyss of creatures and terrains which mostly will never be seen by human eye. Something satisfactory wells up from this deep and nourishes them.

There are extremophiles everywhere, and the adaptations they make to solitariness or small groups are various and strange. By no means all men forced by a sudden change of weather to spend five months' isolation in an Antarctic research station turn out to be either raving or overjoyed when the relief team arrives with thunderous bonhomie and promises of an unbroken year's leave. I was told of one of the last victims of *le bagne* to be released from prison on the Îles du Salut in 1954.[13] He had been there twenty-two years and had benefited from "the cucumber solution" until his fellow convict and lover had died some years previously. The freed man, about to be repatriated to Grenoble, appeared unwilling to leave the island because "he did not know how Tecla would manage without him." Indeed, he begged to remain in Cayenne and to be allowed to go back to his prison island as a caretaker. This was refused. Sadly, he permitted himself to be dressed in a cheap suit and put

[13]My informant was the curator of the museum in Cayenne, *c.* 1972.

aboard a steamer. He took with him a manuscript, "very illegible, a poem of many thousands of lines, all written in pencil on different scraps of paper. I saw it myself, *monsieur*. It was a poem to Tecla, *about* Tecla, and for all I know *by* Tecla. And who was Tecla, you ask? Tecla, *monsieur*, was a gull with one leg. His companion of six years, or so he asked us to believe."

A good example of a highly specialized social group living in extreme circumstances is that of seventeenth-century English pirates in the Caribbean. It was described recently by a scholar who, in default of extensive documentation (for they were not great diarists) elegantly deduced by a series of negative proofs how their lives had to have been. His thesis is that these buccaneers were practically all homosexual and that their piratical activities were sustained by their sexual relationships, much as the Spartans' valor was. He cites as determining factors the generally lenient prevailing attitudes to homosexuality in England at the time and the way in which apprentices were drawn or press-ganged extensively from the bands of boy vagrants ("great flockes of Chyldren") who roamed the country and whose group identification, for their own protection, was exclusive. Furthermore, the population of the British West Indian islands was then almost entirely male, an imbalance enhanced by transportation. "The single certainty is that the only nonsolitary sexual activities available to buccaneers for most of the years they spent in the Caribbean and for almost all of the time they were aboard ship were homosexual."[14] Very few pirates ever married, it seems, and those who did were uniformly unlucky with their women (and vice versa, one would imagine). Blackbeard, William Dampier, Bartholomew Sharp and other pirate captains jealously guarded their favorite boys, while all aboard took advantage of a form of male bonding discreetly named *matelotage*. It is a great pity there is such a dearth of contemporary documents, of poems to Tecla even, although vivid details do emerge. Captain John Avery was known as Long Ben, "not because of his height." Add to all this the occasional bouts of derring-do and the frequent orgies of drinking when every soul aboard from captain to ten-year-old powder monkey was stone drunk, and by contrast to the solidarity of outlaw life afloat, that of "straight" life (in both senses)

[14]B. R. Burg, *Sodomy and the Perception of Evil. English Sea-Rovers in the Seventeenth Century Caribbean* (1983).

ashore would have seemed dreary indeed. It is strange to think that, but for the lack of a few hundred women in the West Indies, piracy might have assumed quite different patterns or even have been suppressed entirely by privateers.

Nomads have a need to wander in a world they understand. Like *Halobates,* many tribes and individuals are very specifically adapted and cannot resist gross change. Take away their habitat, their rovings and solitudes, and they go to pieces. Because of the prevailing cultural pressures in Sulu the Bajau are mainly Islamicized and have become or are becoming Muslim. It is an appropriate religion because Islam contains a statutory requirement for pilgrimage. Perhaps since Islam itself evolved from nomadic cultures it made pilgrimage one of the four chief ritual and moral duties incumbent on all Muslims. In Christianity it is a tradition which has long since faded, except in small groups of hysterics. There is a psychological accuracy in this insistence that a proper life cannot be lived without pilgrimage, a journey, a great excursion and abandoning of town, village, hearth. Only in this way is the unsuspected majesty of the world revealed. Only by traveling in danger and discomfort along arduous forest paths, desert routes and sea lanes may a truth be approached.

~

What is it that a Bajau sees? In one sense, more than any oceanographer; in another, less. In that his vision is peculiar to his living conditions it constructs a set of knowledge about the sea and the world which in the long run must be doomed, although it could be partly resurrected by anyone taking up that life. Much of this vision derives from a literal and daily view of what goes on under the sea. It is informed by the two major characteristics of the tropical waters he inhabits: clarity and phosphorescence. By day the water's transparence declares its emptiness; at night it is filled with blazes of cold fire. This powerfully strikes even those who can talk about phytoplankton.

In the coastal towns and settlements of Sulu the outermost stilt huts of Samal and Bajau may stand 100 yards offshore, extending landward in a dense maze of duckboards and walkways and pilings. Having landed on the rickety wharf it is often impossible to determine exactly where the shoreline is. Gradually the gangplanks between hut and hut sink lower and turn into paths made of embedded lumps of coral the size of babies' heads, damp causeways below floor level with the glint of water on either side. One can often suppose oneself well ashore and then glimpse sumplike pools beneath the huts littered with a floating debris of rotten thatch, splinters of bamboo, boatbuilders' wood shavings, plastic bags, bobbles of excrement, a dead kitten. One would expect these pools to be fetid, miasmic, black. Anywhere else they would be. Yet a muddy bottom is generally visible, often startlingly clear between the bits of flotsam. Perhaps this is simply because the daily tidal creep is just enough to prevent the water stagnating, although any rise and fall is imperceptible.

Rather, it is as though the seawater here contained a particular natural ingredient, *transparency,* which is proof against the cloudy, the dank and the foul. One imagines the water so vibrant that everything in it is strongly polarized. Things are either abundantly alive or else they are whitened bone and shell. There is hardly time for any intermediate stage of decay and it darkens the water not at all. Away from the huts, out over the reefs and shoals, the water is a huge lens focusing light and embalming everything in purest blue glass.

We would say this extreme clarity actually testified to a deficit not of impurities but of nutrients in the waters of the tropics. These seas are relatively stable compared with the colder, stormier oceans. Hence in their euphotic zone the nutrients tend to sink straight down and photosynthesis is correspondingly slower. The deep blue of tropic water is the sea's equivalent to the ocher color of deserts, and likewise signifies an absence. It is hard to believe that cold gray northern seas should be more nourishing than pellucid tropical oceans with their teeming reef life; yet it is so.

After dark, however, these apparently thin waters are rich with the lights they contain. One moonless night I watched some Bajau, scared of lighting a lantern for fear of attracting attention, dive in total darkness for a lost saucepan. The pot had sunk in shallow water, only two or three fathoms, but it still appeared a hopeless task.

I have never seen phosphorescence as bright as on that night. Leaning over the edge of the *bangka* I could follow every move of the searchers below. But the whirligigs of sparks, the flashings and showers of cold fire were at depths which could not be determined. Just as the glints and refractions in the best opals can appear deeper than the thickness of the stone itself or else closer than its surface, so the divers' movements excited discharges of light which were either a few feet away or in a universe beyond. It was vertiginous to gaze down because the view was more what one normally expected to see overhead. On nights as dark as that it is anyway hard to define the horizon, to separate black sky from black sea. Now it was as if the cosmological figures of Sagittarius and Orion had come to life in a firmament beneath the boat. Legendary men outlined in stars swam among clouds of dark matter, galaxies and nebulae swirled in the eddies of their passing. In that moment I could have glimpsed a figure of the Bajau's own believing, the *sama sellang*, or ocean gypsy, who lives in the depths and sometimes leaves his kingdom at night to walk the beaches, a black and shining giant twelve feet tall. He is King of the Fishes, and any Bajau wanting to fish in his domain should have the courtesy and sense to ask his permission first or risk having his boat turn turtle without warning. It was while I was leaning over and watching that one of the outriggers inexplicably sheared off, tipping the *bangka* over and myself into the sea in a sheet of flame. In only a minute or two the Bajau had found their pot on the seabed by the light of their own hands. They emerged with it, laughing, and we all hauled ourselves from the ocean running with greenish fire.

Such are the nights when it is hard for a swimmer to resist heading downward, trailing constellations in a fading dust, and simply go on swimming into the fathomless, sparkling spaces below. It does not feel suicidal in intention, nor like an attack of calenture.[15] Rather, it is more akin to something which might also have played a part in the Bajau's ancient renunciation of land: a momentary expression of yearning for oceanic roots.

~

[15]"Calenture" used to be defined as a tropical shipboard fever whose intolerable burning sometimes made sailors jump into the sea. If this seems overdetermined, a secondary meaning has gradually been allowed to surface to explain the behavior of individuals who, though feverless, may leap overboard without warning and vanish. In this sense calenture becomes a species of mystical rapture, a yearning to blend with the infinite, and has been cited to explain baffling disappearances by lone yachtsmen like Donald Crowhurst and—more recently—the death of the newspaper proprietor Robert Maxwell. If the condition could take a mass-hysterical form it might even throw light on the enigmatic desertion of the *Mary Celeste* in 1872.

It is one of those days when everything we do feels final. The sea is stained with cloud and veined white with its own collapse. It is as if we were having to bid farewell to an entire set of knowledge, to a lifetime's habits. Nothing will come in their place, not epiphany, not even a proper death. Nomads, too, keep track of the seasons, the stately heeling of the stars. So an anxious internal creature (O twentieth century!) is forever flicking over its left wrist in a gesture unknown before the invention of the wristwatch. This modern flip of the knuckles encodes a world of power and—by implication—of powerlessness before a constant anxiety. No doubt there was once an equivalent gesture for a man of the world, perhaps a particular straying of the right hand as toward a sword's pommel, half unconscious, the weapon (no more than an article of dress) forever undrawn, just as the time remains unmemorized: somewhere between a fossil gesture and a social tic.

One need not wear a watch to feel an imaginary arm bend and a wrist turn. We glance at the sun. Even that looks final, seen through the haunted lens of unease. Something is coming. Something is on its way to break things up. We are vaguely prepared for it, unable to settle to anything, as on the day before a journey, knowing it is out of our hands. When it finally arrives it will happen to us, willy-nilly. On a day like this look up at the sun and the clouds. Look at the sea. It is all written there.

Meanwhile, what of the swimmer who lost his boat at the beginning of this book and was left all alone with his panic in the middle of the ocean? That he is here to write the question means the sharks did not get him. Nor pirates, nor fishermen. The engine he heard was another figment. After a long, long time his sight cleared or light rays unkinked and there, no more than eighty yards away, sat the boat. It was solid and unmistakable. It had the air of never having moved, of being practically nailed to the sea, while the swimmer immediately felt his limbs achingly heavy as though he had been on a long and willful excursion and had returned to his senses in the nick of time.

More mysterious still, regaining the boat was like putting on a pair of spectacles, for the low palm-fringed coast became visible exactly where it ought to have been. Ever since that day the swimmer has been unable to account for what happened other than by using a cryptic phrase such as that he fell out of one gaze and into another. He had made the visible a little too hard to see, even though his life depended on it.

Later, he decided the sensation had been less of being lost in the sea, or lost to the sea, than of *the sea's being lost to him*. He was surrounded by water which could have engulfed him; yet at the same time it was a sea which had receded in a way not immediately obvious, taking with it whatever was essential—knowledge, perhaps—for survival. For a long moment there was a boundary fixed around him, an exclusive zone of taint, while perhaps monsters did swim up unseen from the deep, sadly, to gaze at a pair of tiny white limbs cycling high in their skies on the very edge of space. Even had they eaten him the time of their dominion was past. Eventually the legs vanished and the swimmer made off, leaving silvery paddle pocks like fading footprints. The long subsequent journey, of thinking about the sea and the oceans, showed him he was treating them as something which had already been lost.

The gaze of the nineteenth-century explorers and oceanographers was emphatically not the same as our own. They still saw through the eyes of mariners, wary and respectful, as well as with the awed curiosity of intellectual men to whom the ocean was a vast leftover from Creation, a divine challenge which could be met only with the right degree of humility, bravery and methodical amassing of facts. Raw nature was often fearsome. Wild animals were mostly a menace, and even if killing them was not strictly necessary for survival their deaths spoke of a proper relationship between man and nature, sanctioned by Scripture. "It will not be denied, we presume, that animals were created for the use of mankind."[16] In this clear light, landscapes, too, were a threat. Mountains were climbed in the last century as much to tame them as anything else. Nearly impenetrable areas of Africa were explored and named and dragged onto the map, into the domain of "civilized" man (the native gaze of those already living there, and especially the nomadic gaze, did not count). Such was a colonist's view with acquisition on his mind. As for regions like the Antarctic, "Great God," exclaimed Scott, "this is an awful place."

But now, and mostly within the space of the last few decades, this gaze has vanished and modern eyes see an utterly different, less awesome, world. Apart from the fact that nothing a modern oceanographer could discover in the ocean would be likely to precipitate a spiritual crisis, and that technology has made "mariners" in the old sense obsolete while

---

[16]Editorial, *Edinburgh New Philosophical Journal* 2 (1860), p. 283.

removing a good deal of science from the domain of humility and bravery and dumping it instead into the frankly humdrum—apart from all this, we can now only ever look at the natural world with the anxious gaze of conservationists. It is all falling away, becoming lost to us. No longer are the wilds and unoccupied spaces of Earth things to be tackled and subjugated. Rather, they are to be cherished and protected because on them depends our survival. The view through our sunglasses and snow goggles and diving masks is suddenly of last things. The midnight chorus of a reef filling our ears as we clutch in darkness twenty feet down is the voice of the sea rehearsing its own eschatology.

Conrad dated the beginning of the end of the old sea from the building of the Suez Canal, completed in 1869.

Then a great pall of smoke sent out by countless steamboats was spread over the restless mirror of the Infinite. The hand of the engineer tore down the veil of the terrible beauty in order that greedy and faithless landlubbers might pocket dividends. The mystery was destroyed. . . . The sea of today is a used-up drudge, wrinkled and defaced by the churned-up wakes of brutal propellers, robbed of the enslaving charm of its vastness, stripped of its beauty, of its mystery and its promise.[17]

That was first published in 1896, and its jeremiad tone is similar to many other laments over industrialization. A century later, only vestiges remain of the sea Conrad described, let alone of the one he mourned.

Yet even this is not quite the reason for the swimmer's conviction that the sea was lost to him. The oceans have long been, and will long be, subjected to ruthless exploitation and even, in places, to ruin. It is not really the sea which is in recession, though, but wildness itself. Wildness is everywhere but it can no longer be seen; and its apparent vanishing is a direct consequence of the new conservationist gaze. "The Wild" is nowadays a concept ringing with the overtones of patronage, of collections by schoolchildren on its behalf. The present generation is as much contaminated by its own reverential and placatory attitude as the older was by domination. There is something ignoble about it, compounded as it is of urban sentimentalism, virtuous concern and sheer panic at having irrevocably fouled the nest while so comfortably lining it. Above

[17]Joseph Conrad, *An Outcast of the Islands* (1896), beginning of Chapter 2.

all, the self-interest shows through. Luckily, there is a chasm properly and forever fixed between the nonhuman and the humanist biospheres, between wildness and *caring*. It is seldom visible to modern eyes. Virtue and the wild share no common universe.[18]

If the sea always was a rich source of melancholy, there is in any case a new melancholy to go with the new gaze. Conservation is only ever a rearguard action, fought from a position of loss. It is ultimately unwinnable, and not least because there are no recorded victories over population increase, nor over the grander strategies of genetic behavior such as the laws of demand, political expediency, sheer truancy and a refusal to relinquish a standard of living once it has been attained. There can only be stalemates, holding actions and truces uneasily policed. A few affecting species will be saved, a few million hectares of forest, a few tribes of Indians; but the world will never return to how it was when this sentence was written, still less to how it was when reader and writer were born. This has always been true and will continue to be so. The mistake is to extend this sequence backward in time and imagine it leads to a lost paradise. It is a safe bet that as soon as the earliest protohominid could think, it invented a legend to account for its sense of loss.

But the *swimmer* . . .

The hypocrite swimmer has himself lost all interest in these arguments. He is intently reaching the shore in his little boat, paddling carefully among familiar corals, following the narrow channel in toward the beach. He has already forgotten his panic and is merely tired as from a long journey. Cool in the bilges lie half a dozen mackerel: two for supper, two to smoke for tomorrow and two to give away. He looks up

---

[18]It is a wishful belief on the part of many that in some mystical way animals can appreciate human social values such as "goodness." "They know," is the usual sage assurance, "they can tell." On the contrary, all evidence suggests that animals have not the least interest in morals, or else they are remarkably undiscriminating. There is an account in Browne and Tullett's biography, *Bernard Spilsbury,* of the murderer Patrick Mahon. In the 1920s, in a house on a desolate stretch of shingle near Eastbourne, Mahon cut up and rendered down in a caldron the *disjecta membra* of Emily Kaye, the girl he had made pregnant. Mahon, "like St. Francis, whom he resembled in no other way, had a remarkable influence over animals. Those who like to think that animals know good people from bad will be distressed to learn that Mahon had only to whistle and birds came to him, and that dogs and cats deserted their masters and mistresses to follow him home." One assumes the explanation was something like pheromones. Likewise, animals flocked to St. Francis not because he was a saint but because they happened to like the smell of his glands.

as the prow grinds into the sand. There in the palms' ragged shade is his lopsided hut, there the tangle of thorn shrubs concealing a mahogany-colored brackish lagoon, in the distance the spit of mangroves walking on water. He sees it all not through the eyes of an oceanographer, still less of a conservationist. Only in a nomad's or a wanderer's gaze is the sea not lost to him, nor any less wild. So affectionately does the scene bound toward him and leap into his eye that he knows this private way of looking reveals a landscape he must have inherited, or which was somehow fixed for him as a child, before he ever saw it for the first time.

For this is his ocean, and at last he knows he has always seen it thus, toward the end of afternoon: the great white clouds heaping themselves out of nothing against the blue, their tall reflections falling on a glassy sea whose tide lies stilled at low. Reef tops knobble the surface, the kelps and grasses float as rough brown patches among which the white clouds lie in fragments. Children stand a hundred yards out, up to their ankles, legs angular as wading birds', filling coconut shells and tins with winkles. They dabble among the white clouds. Clear voices drift ashore, tatters of a heedless present.

It is the moment of being aghast at the sad miracle of having condensed from nothing, of watching white clouds, of dispersing again. But how beautiful it is; and how pierced by it we always are as it leaps through us, and leaping, vanishes.

3. STEAMERS/STREAMERS

The Sulu Archipelago is a good example of a place which must be reached by boat if it is ever to be seen. Only a boat, as opposed to an aircraft, will put the traveler within its coordinates. There is always some risk of attack by pirates or of going down in a vessel like the *Doña Marilyn* and it is important to court that risk. Besides, the cramped, hot, vomity approach through a sea sprinkled to the horizon with small islands is the correct one. It is necessary to wake at dawn on a folding canvas deck bed jammed between its neighbors like a stretcher in a busy field hospital, face clammy with salt and dew and whipped by strands of a stranger's hair. Out of that fitful, blurred sleep an island has emerged on a turquoise sea and those whose destination it is begin to stir, waking their children, pulling their belongings together. This slow, oneiric approach must be observed. No place ever quite survives the wrong landfall.

If one wished to formulate a First Law of Travel, it might be that the mode of travel determines the place reached. To take an example: The Korea one reaches by cycling from Hamburg (which could be done without ever once glimpsing the sea) is altogether a different place from the Korea reached by flying from Heathrow or Kennedy. The people are different, they speak a different language and have a far greater knowledge of bicycles. This principle holds good for Sulu, too, and is likewise dependent on adopting the right pace. All those passenger ships, more or less unseaworthy, which still ply the networks of archipelagic routes are simply ferries, shuttling people around within a country. They make one lament the passing of international sea travel. The sundry boredoms and discomforts of passenger liners were as nothing compared to those of the aircraft which have replaced them. There is a new generation of cruise liners, it is true, but they scarcely touch the argument. Cruise liners are not going anywhere, so they function more or less as hotels, with the novelty that they float and move. They are as aimless as the pleasure they pursue, that classic wild-goose chase. As was memorably said in a radio program about the hijacked *Achille Lauro,* however luxurious a modern cruise liner "it's really no more than a velvet-lined prison hulk."

Just as modes of travel affect destinations, so do they change our farewells and the very nature of separation. Airport terminals swallow up friends and lovers in a way docks never did. The drabness of quays is not to be compared with the squalor of concourses. Even if we bother to wave foolishly from the terminal roof as an anonymous metal toy half a mile away disappears within seconds into gray overcast, we can never be sure it was the right one. In any case, part of the dispassion of these aerial bus journeys lies in their being so swift. It now takes effort and care for two people to be distant from one another by much more than twenty-four hours.

Ships, on the other hand, carry with them the solemnity of long separations, perhaps of lifetimes. Stately valedictions echoed through people's lives until thirty or so years ago. Ships are individual as one of a fleet of Boeings never could be, even though an aircraft may carry a name (often more or less geared to the tourist age: *Loch Ness, Val d'Aosta,* and so on). A ship *is* her name, right to the bottom and beyond, connoting a moment of history as much as a vague locus on the seabed (*Titanic, Lusitania, Andrea Doria*). Aircraft when they crash shed their names along with their wings. They become "ill-fated Flight 307"; or else Pan Am flight 103 becomes "Lockerbie," an entire nexus of loss reduced to a point of impact.

The departure of a ship is slow, celebratory, mournful. It gives time to think and the proper space in which to let fall one's lesser salt into the greater below. Something of moment is happening, part of whose subtext is a fear or resentment of the sea as the agent of long absences, slow letters and terrible news. Whoever they are, down at the docks one windy afternoon—friends, lovers, siblings—they are already separated. There are those on the quay and those already on board, though both are watching. The ship is about to sail. Gangways are lowered, ropes cast off. Heavy nooses splash into the slot of oily scum between truck-tire fenders and iron cliff. Cries go up. The siren's blare, of such low frequency it shakes the stomach and jars loose fresh tears, sounds once, twice. Yet an illusion of unparting is preserved by the streamers, cheerful strips of paper sagging and twirling between the thousand pairs of widely separated hands.

Over the whole scene hovers loss looking for somewhere to settle. Is it in the already spoken good-bye? In the last touch of bodies? In the cries of the gulls? Or does it now pulse along that thin paper nerve? It parts; they part. Yet still they remain visible to each other while loss fills up the space opening between them, stretching out between ship and shore, between hull and headland, dot and smudge, before spreading across the face of the globe. But of course it had been there in the train on the way to the port. And before, in the careful packing of suitcases. And before that.

And after? Here again, air travel offers no comfort because its speed runs departure into arrival, leaching out their difference, blurring them into a hectic placelessness. We cannot tell what to think. It is too brutal, facing the ordeal of a dawn landing on a strange continent with the scents of leavetaking still in our clothes. We have walked the streets of an Asian city with the fresh scratches of a cat in Oxfordshire on our hands. Such confusions make unreal both cat and city, and leave us feeling we can never properly come to grips with anything.

Traveling long distances by sea, on the other hand, gives us time. Travel is like death in that it requires mourning. The light melancholy of watching a coastline recede is a necessary observance. We join in with shipboard life just as soon as we wish, and not before. Otherwise we write in our cabin or spend hours watching the wake of our own passage. The caves sucked into the water's surface by the turning of invisible propellers—each subtly different, each marbling a dissipating track which stretches back, an elastic streamer—become hypnotic. They set us adrift on inward voyages where we barely have enough sarcastic energy left to stop ourselves seeing our frail barks upon the vasty deep as paradigmatic. Such time, such long hovering on the edge of banality, is powerfully restorative. By the time approaching land is announced we are free to be excited. Later, it seems to us that only by having breathed the salt air of loss for long enough are we able to make a properly carefree disembarkation. We have adjusted. Our biological clocks are reset, our homoiothermal balance has altered with the latitude, our internal maps—whose every nautical mile has been *felt* as traveled— make sense. Behind us the ocean is crisscrossed with thousands upon thousands of multicolored streamers, a planet festooned with farewells.

INDEX

ABOUT THE AUTHOR

JAMES HAMILTON-PATERSON was educated at Oxford, where he won the Newdigate Prize. In addition to extensive journalism for *The Sunday Times,* the *Times Literary Supplement* and the *New Statesman,* he has published volumes of poetry and short stories. The nonfiction work *Playing with Water* was followed by his first novel, *Gerontius,* which won the Whitbread Prize in 1989, and *The Bell-Boy.* He is a Fellow of the Royal Geographical Society and lives in Italy and the Far East.

ABOUT THE TYPE

This book was set in Galliard, a typeface designed by Matthew Carter for the Mergenthaler Linotype Company in 1978. Galliard is based on the sixteenth-century typefaces of Robert Granjon, which give it classic lines yet interject a contempory look.